BOLYAI SOCIETY
MATHEMATICAL STUDIES

16

BOLYAI SOCIETY MATHEMATICAL STUDIES

1. **Combinatorics, Paul Erdős is Eighty, Vol. 1**
   D. Miklós, V.T. Sós, T. Szőnyi (Eds.)
2. **Combinatorics, Paul Erdős is Eighty, Vol. 2**
   D. Miklós, V.T. Sós, T. Szőnyi (Eds.)
3. **Extremal Problems for Finite Sets**
   P. Frankl, Z. Füredi, G. Katona, D. Miklós (Eds.)
4. **Topology with Applications**
   A. Császár (Ed.)
5. **Approximation Theory and Function Series**
   P. Vértesi, L. Leindler, Sz. Révész, J. Szabados, V. Totik (Eds.)
6. **Intuitive Geometry**
   I. Bárány, K. Böröczky (Eds.)
7. **Graph Theory and Combinatorial Biology**
   L. Lovász, A. Gyárfás, G. Katona, A. Recski (Eds.)
8. **Low Dimensional Topology**
   K. Böröczky, Jr., W. Neumann, A. Stipsicz (Eds.)
9. **Random Walks**
   P. Révész, B. Tóth (Eds.)
10. **Contemporary Combinatorics**
    B. Bollobás (Ed.)
11. **Paul Erdős and His Mathematics I+II**
    G. Halász, L. Lovász, M. Simonovits, V. T. Sós (Eds.)
12. **Higher Dimensional Varieties and Rational Points**
    K. Böröczky, Jr., J. Kollár, T. Szamuely (Eds.)
13. **Surgery on Contact 3-Manifolds and Stein Surfaces**
    B. Ozbagci, A. I. Stipsicz
14. **A Panorama of Hungarian Mathematics in the Twentieth Century, Vol. 1**
    J. Horváth (Ed.)
15. **More Sets, Graphs and Numbers**
    E. Győri, G. Katona, L. Lovász (Eds.)
16. **Entropy, Search, Complexity**
    I. Csiszár, G. Katona, G. Tardos (Eds.)

Imre Csiszár
Gyula O. H. Katona
Gábor Tardos (Eds.)

Entropy, Search, Complexity

Springer
JÁNOS BOLYAI MATHEMATICAL SOCIETY

Imre Csiszár
Hungarian Academy of Sciences,
Alfréd Rényi Institute of Mathematics
Reáltanoda u. 13–15
1053 Budapest, Hungary
E-mail: csiszar@renyi.hu

Gyula O. H. Katona
Hungarian Academy of Sciences,
Alfréd Rényi Institute of Mathematics
Reáltanoda u. 13–15
1053 Budapest, Hungary
E-mail: ohkatona@renyi.hu

Gábor Tardos
Hungarian Academy of Sciences,
Alfréd Rényi Institute of Mathematics
Reáltanoda u. 13–15
1053 Budapest, Hungary
E-mail: tardos@renyi.hu

Managing Editor
Gábor Wiener
Budapest University of Technology
and Economics
Pázmány Péter sétány 1/D
1117 Budapest, Hungary
E-mail: wiener@cs.bme.hu

Mathematics Subject Classification (2000): 94A15, 90B40, 68P10

ISSN 1217-4696
ISBN 978-3-642-06899-7 e-ISBN 978-3-540-32777-6

Springer is a part of Springer Science+Business Media
springer.com

Softcover reprint of the hardcover 1st edition 2007

Cover design: Erich Kirchner, Heidelberg

Printed on acid-free paper 44/3142/db – 5 4 3 2 1 0

Contents

Contents 5
Preface 7
M. Aigner: Two Colors and More 9
C. Deppe: Coding with Feedback and Searching with Lies 27
A. G. D'yachkov, A. J. Macula and P. A. Vilenkin: Nonadaptive and Trivial Two-Stage Group Testing with Error-Correcting d^e-Disjunct Inclusion Matrices 71
S. Ghosh, T. Shirakura and J. N. Srivastava: Model Identification Using Search Linear Models and Search Designs 85
P. Harremoës: Information Topologies with Applications 113
M. Keane: Reinforced Random Walk 151
D. Petz: Quantum Source Coding and Data Compression 159
F. Topsøe: Information Theory at the Service of Science 179
P. Vitányi: Analysis of Sorting Algorithms by Kolmogorov Complexity 209
G. Wiener: Recognition Problems in Combinatorial Search 233

Preface

The present volume is a collection of survey papers in the fields given in the title. They summarize the latest developments in their respective areas. More than half of the papers belong to search theory which lies on the borderline of mathematics and computer science, information theory and combinatorics, respectively. The volume is slightly related to the twin conferences "Search And Communication Complexity" and "Information Theory In Mathematics" held at Balatonlelle, Hungary in 2000. These conferences led us to believe that there is a need for such a collection of papers.

The paper written by Martin Aigner starts with the following relatively new search problem. Given n boolean variables as input one has to find one of them whose value is in majority. The goal is to minimize the number of tests needed for this where one test is to compare two input variables for equality. The paper surveys the large set of problems and results which grew out of this one.

In the traditional search model an unknown element is sought in a finite set, based on the information that the unknown element is or is not in some (asked) subsets. A variant is when a $0, 1$ function is given on the underlying set, and only the values of this function at the unknown element x is sought rather than x itself. This is called the recognition problem. Gábor Wiener's paper shows that the recognition problem actually includes the problem of two-party, deterministic communication complexity. Using this novel observation it unifies and surveys results in both theories.

The theory of search with lies, or the Ulam–Rényi game is an exciting area with many applications. The paper of Christian Deppe gives a complete survey on search problems obtained by allowing lies, that is, when the answer on the question "is the unknown element x in the subset A?" can be wrong, but the number of lies is limited.

In linear statistics the influence of certain statistical parameters (factors) and their combinations are to be determined. Traditionally it was supposed that one knows beforehand a few important variables and we assume that any combination of the remaining variables have negligible influence. A 20 year old theory does not make this assumption. Instead, the experiments have two simultaneous goals: 1. determine which combinations have a non-negligible influence, and 2. find the influence of these combinations. This is why this setting is a generalization of the search problems. The paper of S. Ghosh, T. Shirakura and J. N. Srivastava gives a strong survey of the results in this theory. Let us mention the theory was founded by the last of these authors.

The basic problem of sorting is to find the natural order of a set of integers by pairwise comparisons. It is easy to see that this also fits in the search model: the underlying set is the set of all permutations, the unknown element is the actual permutation determined by the natural order. A relatively new development of the theory that Kolmogorov's complexity can be used in proving bounds in sorting problems. This new theory is surveyed here in the paper of Paul Vitányi.

The paper of D'yachkov, Macula and Vilenkin contains some new results in the area of non-adaptive search with more than one unknowns. However it adds an extensive literature in the given area which helps the reader to obtain a good view.

Flemming Topsøe surveys the situations where information theory can be used. It has a novel attitude: the situations are treated as problems of games. 60 references help the reader to study the details.

Dénes Petz gives a survey of the very modern area of Quantum Source Coding.

The work written by Michael Keane is not a survey. It poses "only" an exciting new problem that is both very natural and easy to formulate. On the other hand the paper makes it clear that it is actually a starting point of a class of difficult problems. The conclusion is a brief description of the known results.

Peter Harremoës's paper introduces new topologies on probability distributions, that is, information theoretical divergencies. Since they are compared with the traditional divergencies (entropies), the paper contains a good survey of these information theoretical concepts. The importance of the new concepts are justified by theorems, too.

The editors

BOLYAI SOCIETY
MATHEMATICAL STUDIES, 16

Entropy, Search, Complexity, pp. 9–26.

Two Colors and More

MARTIN AIGNER

Suppose we are given n balls colored with two colors. How many color-comparisons are needed to produce a ball of the majority color? The answer (first given by Saks and Werman) is $M(n) = n - B(n)$, where $B(n)$ is the number of 1's in the binary representation of n. We consider in this paper several generalizations and variants of the majority problem such as producing a k-majority ball, determining the color status of all balls, arbitrarily many colors, the plurality problem, and the closely related liar problem.

1. The majority problem

Suppose we are given n balls colored with two colors, and two players Paul and Carole playing the following game. At any stage of the game Paul chooses two balls x and y and asks whether they are of the same color, whereupon Carole answers "yes" or "no". The game ends when Paul either produces a ball z of the majority color (meaning that the number of balls colored like z exceeds the other color), or when Paul states that there is no majority. Of course, the latter case can only occur when n is even. How many questions $L(n)$ does Paul have to ask in the worst case?

This problem was first solved by Saks and Werman [11] and later by Alonso, Reingold, Schott [4] and Wiener [12] using different methods. The answer is

$$L(n) = n - B(n), \tag{1}$$

where $B(n)$ is the number of 1's in the binary representation of n. Alonso, Reingold and Schott [4] also gave the solution for the average case.

As a warm-up let us see how Paul finds an algorithm that uses no more than $n - B(n)$ questions. The data structure during the game is a list of buckets $B_1, \ldots, B_s$ and a dump D

$$\begin{array}{ccccc} \boxed{2^{a_1}} & \boxed{2^{a_2}} & \cdots & \boxed{2^{a_s}} & \boxed{} \\ B_1 & B_2 & & B_s & D \end{array}$$

where the balls in each bucket are colored alike, always numbering a power of 2. Thus, initially, there are n buckets each containing one ball with the dump empty. For the next test Paul chooses two buckets B_i, B_j with $a_i = a_j$ and compares balls from B_i and B_j. If the answer is "yes", he merges the buckets into one (of new size 2^{a_i+1}), otherwise he empties both buckets into the dump. Hence D contains at any stage an equal number of either color.

The algorithm stops when either all buckets have different sizes $2^{b_1} > 2^{b_2} > \cdots > 2^{b_t}$, or when all balls are in the dump. In the first case the size 2^{b_1} of the largest bucket exceeds $2^{b_2} + \cdots + 2^{b_t}$, and we conclude that B_1 contains the majority color balls. In the other case there is no majority. Hence with either alternative the game is finished.

It remains to compute the number L of questions. By induction it is clear that Paul needs $2^{b_i} - 1$ questions to produce a bucket of size 2^{b_i}. Similarly, when he throws two buckets of equal size $2^{c_i - 1}$ into D then he has asked $2^{c_i} - 1$ questions. Hence

$$L \leq (2^{b_1} + \cdots + 2^{b_t} - t) + (2^{c_1} + \cdots + 2^{c_r} - r)$$

where

$$n = 2^{b_1} + \cdots + 2^{b_t} + 2^{c_1} + \cdots + 2^{c_r}.$$

Since obviously $t + r \geq B(n)$, we obtain

$$L \leq n - B(n).$$

It is the aim of this paper to present a survey of several natural generalizations and variants of the majority problem, including a number of open questions. Only a few proofs will be given in full detail, the emphasis being on the common ideas for this appealing part of combinatorial search. For the general background the reader may consult the books by Aigner [1] or Du–Hwang [7].

2. First Generalization: Determining a k-majority

We are again given n balls colored with two colors, and a threshold $k > \frac{n}{2}$. In the (n,k)-majority game Paul must exhibit a *k-majority* ball z (that is, there are at least k balls colored like z), or declare that there is no k-majority. Let us denote by $L(n,k)$ the number of questions in the worst case. Hence the original problem calls for $L(n) = L\big(n, \lfloor \frac{n}{2} \rfloor + 1\big)$. Note that we always have

$$L(n,k) \leq n-1, \tag{2}$$

since Paul may compare a fixed ball to all the others. Indeed, with this procedure Paul determines the full color partition. We will return to this aspect in the last section.

It is convenient to rephrase the game as follows (see [5, 11]). Draw an edge between x and y when x and y are compared. Suppose at a certain stage of the game $C_1, \ldots, C_s$ are the components. Within each C_i the color classes and hence their sizes a_i and b_i are known from the answers. Denote by $m_i = |a_i - b_i| \geq 0$ the difference between the majority and minority number. Hence the stage can be completely described by the *state vector*

$$M = (m_1, \ldots, m_s).$$

If Paul compares next a majority ball of C_i with a majority ball of C_j, then the answer "yes" results in

$$M_{ij}^{+} = (m_1, \ldots, m_i + m_j, \ldots, m_s)$$

and the answer "no" in

$$M_{ij}^{-} = (m_1, \ldots, |m_i - m_j|, \ldots, m_s)$$

with m_i, m_j deleted in both cases.

We note four things:

a) The initial state is $M_0 = (1, \ldots, 1)$ with n 1's.

b) The number of questions asked up to $M = (m_1, \ldots, m_s)$ equals $n - s$.

c) If $M = (m_1, \ldots, m_s)$ is a state, then

$$\sum(M) := \sum_{i=1}^{s} m_i \equiv n \pmod{2}. \tag{3}$$

This holds since $\sum(M_0) = n$ and either answer $m_i + m_j$, $|m_i - m_j|$ does not change the parity of the sum.

d) Let $M = (m_1, \ldots, m_s)$ be a state. Since $2k - n = k - (n-k)$ is the critical difference between majority and minority, we find that the majority color in C_i must be a k-majority if $m_i \geq \sum_{j \neq i} m_j + 2k - n$. On the other hand, if $m_i < \sum_{j \neq i} m_j + 2k - n$, then Paul cannot be sure about the status of the colors in C_i. Similarly as long as $m_1 + \cdots + m_s \geq 2k - n$, a k-majority is still possible, whereas $m_1 + \cdots + m_s < 2k - n$ implies that a k-majority does not exist.

Taking the parity condition (3) into account we can therefore state for $M = (m_1 \geq \cdots \geq m_t)$:

$$(3) \qquad M \text{ is not terminal} \iff 2m_1 \leq \sum(M) + 2k - n - 2 \quad \text{and} \quad \sum(M) \geq 2k - n$$

$$(4) \qquad M \text{ is terminal} \iff 2m_1 \geq \sum(M) + 2k - n \quad \text{or} \quad \sum(M) \leq 2k - n - 2.$$

For a state $M = (m_1, \ldots, m_s)$ denote by $V(M)$ the size t of the terminal multi-set when both players perform optimally. Since the number of questions at the end is $n - V(M)$, Paul wants to maximize $V(M)$ and Carole wants to minimize it. It follows that

$$(5) \qquad V(M) = \max_{i,j} \min\left(V(M_{ij}^+), V(M_{ij}^-)\right).$$

Given $M = (m_1, \ldots, m_s)$ we say that $m_h : m_\ell$ is an *optimal choice* for Paul if

$$(6) \qquad V(M) = \min\left(V(M_{h\ell}^+), V(M_{h\ell}^-)\right) \geq \min_{ij \neq h\ell}\left(V(M_{ij}^+), V(M_{ij}^-)\right).$$

Clearly, an optimal choice never involves a number $m_i = 0$, since $m_i = 0$ does not change the relations in (4).

Theorem 1 [2]. *We have*

$$(7) \qquad M(n,k) \geq n - 1 - p \quad \textit{where} \quad 2^p \parallel \binom{n-1}{k-1},$$

that is, 2^p is the highest power of 2 dividing $\binom{n-1}{k-1}$.

Proof. We proceed along the lines of the Saks–Werman argument in [11] and explain it for completeness. Suppose we can find a function $\phi(M)$ such that for any state M

i) $\phi(M) \geq V(M)$, if M is terminal

ii) $\phi(M) \geq \min\left(\phi(M_{ij}^+), \phi(M_{ij}^-)\right)$ for any i, j.

Then $\phi(M) \geq V(M)$ for all M, and thus

$$n - V(M) \geq n - \phi(M) \quad \text{for all } M. \tag{8}$$

Indeed, if M is not terminal, then for an optimal choice $m_i : m_j$ of M we conclude by induction and ii)

$$\phi(M) \geq \min\left(\phi(M_{ij}^+), \phi(M_{ij}^-)\right) \geq \min\left(V(M_{ij}^+), V(M_{ij}^-)\right) = V(M).$$

For a state $M = (m_1, \ldots, m_s)$ call $I \subseteq \{1, \ldots, s\}$ *big* if $\sum\limits_{i \in I} m_i \geq \sum\limits_{j \notin I} m_j + 2k - n$, and let $f_M(x) = \sum\limits_{I \text{ big}} x^{m_I}$, $m_I = \sum\limits_{i \in I} m_i$, be the generating function of the big sets. For an integer m denote by $P(m)$ the largest power of 2 dividing m, that is $2^{P(m)} \parallel m$, with $P(0) = \infty$. Now we define

$$\phi(M) = 1 + P\big(f_M(-1)\big),$$

and verify i) and ii).

Suppose $M = (m_1 \geq \cdots \geq m_t)$ is terminal with $m_1 \geq \sum\limits_{j=2}^{t} m_j + 2k - n$. Then the big sets are precisely those sets containing 1. Hence

$$f_M(x) = x^{m_1}(1 + x^{m_2}) \ldots (1 + x^{m_t})$$
$$f_M(-1) = (-1)^{m_1}\big(1 + (-1)^{m_2}\big) \ldots \big(1 + (-1)^{m_t}\big).$$

Now either $f_M(-1) = 0$ and thus $\phi(M) = \infty > t = V(M)$, or $\left|f_M(-1)\right| = 2^{t-1}$, in which case we obtain $\phi(M) = t = V(M)$. When there is no k-majority, then there are no big sets, $f_M(x) = 0$, and therefore $\phi(M) = \infty$.

To verify condition ii), suppose $M = (m_1, \ldots, m_s)$ and $M^+ = M_{ij}^+$, $M^- = M_{ij}^-$ with $m_i \geq m_j$. An easy case analysis shows

$$f_M(x) = f_{M^+}(x) + x^{m_j} f_{M^-}(x).$$

Now, $P(a+b) \geq \min\left(P(a), P(b)\right)$, and we infer

$$\begin{aligned}\phi(M) &= 1 + P\big(f_M(-1)\big) = 1 + P\big(f_{M^+}(-1) + (-1)^{m_j} f_{M^-}(-1)\big)\\ &\geq 1 + \min\left(P\big(f_{M^+}(-1)\big), P\big(f_{M^-}(-1)\big)\right)\\ &= \min\left(\phi(M^+), \phi(M^-)\right).\end{aligned}$$

To end the proof consider the initial state $M_0 = (1, \ldots, 1)$. Clearly, the big sets are those containing at least k indices, whence

$$f_{M_0}(-1) = \sum_{j=k}^{n} \binom{n}{j} (-1)^j = (-1)^k \binom{n-1}{k-1}.$$

This yields $\phi(M_0) = 1 + p$ with $2^p \parallel \binom{n-1}{k-1}$, and thus

$$L(n,k) = n - V(M_0) \geq n - \phi(M_0) = n - 1 - p. \qquad \blacksquare$$

Remark 1. It is an elementary fact of number theory that for binomial coefficients: $P\left(\binom{a}{b}\right) = B(b) + B(a-b) - B(a)$, where $B(m)$ denotes as before the number of 1's in the binary representation of m. Hence

$$L(n,k) \geq n - 1 - B(k-1) - B(n-k) + B(n-1).$$

In the ordinary majority game we have $n = 2m+1$, $k = m+1$ or $n = 2m$, $k = m+1$. In the odd case

$$\begin{aligned}L(n) &\geq n - 1 - B(m) - B(m) + B(2m)\\ &= n - 1 - B(2m) = n - B(n),\end{aligned}$$

since $B(2m) = B(m)$, $B(2m+1) = B(2m) + 1$. In the even case we obtain similarly

$$\begin{aligned}L(n) &\geq n - 1 - B(m) - B(m-1) + B(2m-1)\\ &= n - 1 - B(m) + 1 = n - B(n),\end{aligned}$$

since $B(2m-1) = B(m-1) + 1$, and the equality $L(n) = n - B(n)$ is established.

Remark 2. The theorem implies $L(n,k) = n-1$ whenever $\binom{n-1}{k-1}$ is odd. For example $L(2^m, k) = 2^m - 1$ for any k, since it is well-known that $\binom{2^m-1}{k-1}$ is always odd. Conversely, it can be shown that $L(n,k) = n-1$ implies that $\binom{n-1}{k-1}$ is odd.

Remark 3. The bound of Theorem 1 does not always give the correct value. The smallest example is $n = 9$, $k = 6$. Here $\binom{n-1}{k-1} = 56$, hence $L(9,6) \geq 8 - 3 = 5$, whereas the true value is $L(9,6) = 7$.

To further study $L(n,k)$ the following inequalities are derived in [2] using the function $V(M)$:

Proposition 1. *We have*

a) $L(n,k) \geq L(n-1,k) \quad (2k > n)$

b) $L(n,k) \geq L(n-1,k-1) \quad (2k > n+1)$

c) $L(n,k) \geq L(n-2,k-1) + 1 \quad (2k > n)$.

A general formula for $L(n,k)$ is not known, but we can derive a recursive expression for $L(n,k)$ under the plausible assumption:

(H) *There is always an optimal algorithm for Paul which makes $\lfloor \frac{n}{2} \rfloor$ disjoint comparisons first.*

Theorem 2. *Under assumption* (H)

$$L(n,k) = \begin{cases} L\left(\frac{n}{2}, \frac{k}{2}\right) + \frac{n}{2} & n \equiv 0,\ k \equiv 0 \pmod 2 \\ L\left(\frac{n-1}{2}, \frac{k}{2}\right) + \frac{n-1}{2} & n \equiv 1,\ k \equiv 0 \pmod 2 \\ L\left(\frac{n}{2}, \frac{k+1}{2}\right) + \frac{n}{2} & n \equiv 0,\ k \equiv 1 \pmod 2 \\ L\left(\frac{n+1}{2}, \frac{k+1}{2}\right) + \frac{n-1}{2} & n = 1,\ k \equiv 1 \pmod 2 \end{cases}$$

with $L(1,1) = 0$.

The solution of the recursion in Theorem 2 is given in the following result.

Proposition 2. *Let* $n = 2^{a_1} + 2^{a_2} + \cdots + 2^{a_t}$, $a_1 < \cdots < a_t$, *and* $k = 2^{b_0} + 2^{b_1} + \cdots + 2^{b_s}$, $b_0 < \cdots < b_s$, *be the binary representations of* n *and* k, $2k > n$. *Set* $A = \{a_j : a_j \geq b_0\}$, $B = \{b_1, \ldots, b_s\}$. Then under the hypothesis *(H)* $L(n,k)$ is given by

$$L(n,k) = n - r,$$

where

$$r = [\#i : a_i \leq b_0] + [\#j : a_j > b_0 \text{ such that if } b_0 \leq m < a_j, \\ m \in A \setminus B, \text{ then there exists } p \in B \setminus A \\ \text{with } m < p < a_j].$$

Example. Consider $n = 55 = 2^0 + 2^1 + 2^2 + 2^4 + 2^5$, $k = 44 = 2^2 + 2^3 + 2^5$. Then $b_0 = 2$, $A = \{2, 4, 5\}$, $B = \{3, 5\}$. Here $r = 3 + 1 = 4$ since $a_4 = 4$ satisfies the second condition, but $a_5 = 5$ does not. Hence $L(55, 44) = 51$.

Conjecture 1. *Proposition* 2 *gives the true value of* $L(n, k)$ *for all* n *and* k.

3. Second generalization: more colors

The next generalization immediately comes to mind: Instead of two colors we consider c colors, with $2 \leq c \leq n$. Let us denote by $L_c(n, k)$ the worst-case complexity to produce a k-majority ball (or state that there is no k-majority), when the balls are colored with up to c colors. Thus $L(n, k) = L_2(n, k)$.

The following chain of inequalities is obvious:

$$L_2(n, k) \leq L_3(n, k) \leq \cdots \leq L_n(n, k). \tag{9}$$

Fisher and Salzberg determined $L_n\left(n, \left\lfloor \frac{n}{2} \right\rfloor + 1\right)$, that is the length of the ordinary majority game in the presence of n colors [9]:

$$L_n\left(n, \left\lfloor \frac{n}{2} \right\rfloor + 1\right) = \left\lceil \frac{3n}{2} \right\rceil - 2.$$

The general result reads as follows:

Theorem 3. *We have*

$$L_n(n, k) = 2n - k - 1. \tag{10}$$

Proof. We just give the upper bound using a similar idea as in [9]. Thus we have to exhibit an algorithm that uses no more than

$$L \leq 2n - k - 1 \tag{11}$$

questions. For $k = n$ this is clear from (2). For $k = \frac{n+1}{2}$ we have the Fisher–Salzberg result. Setting $d = 2k - n$ we may thus assume $1 < d < n$.

Phase 1. Order the balls $x_1, x_2, \ldots, x_n$, and compare the x_i's one after the other. We set up a dynamic list L and a reservoir R. Initially $L = \{x_1\}$, $R = \emptyset$. Suppose that before the ball x_j is compared, the list is $L = y_1 y_2 \ldots y_s$ with reservoir R. Now compare x_j with the last ball y_s of the list. If they have the same color, put x_j into R. Otherwise, enlarge L by moving x_j to the end of L and putting a ball z of R behind x_j (in case $R \neq \emptyset$):

$$L \to L x_j z, \quad R \to R \setminus \{z\} \quad (\text{if } R \neq \emptyset).$$

The following facts are immediate:

a) All balls in R have the same color which is equal to the color of the last ball in L.

b) Neighboring balls in L have different colors.

c) At the end of phase 1 (that is, after $n-1$ tests) the only possible k-majority color is that of the last ball in L (or of R).

d) If there is a k-majority, then the (unknown) difference

$$\delta = \# \text{ majority balls} - \# \text{ rest} \tag{12}$$

must satisfy $\delta \geq d$.

Phase 2. Let $L = \ldots y_3 y_2 y_1 b a$, $|R| = r$. If $r \geq n-2$, then a is a majority ball, and we are through. Hence we may assume $r < n-2$. We use δ as in (13) and denote by ρ the (dynamic) size of R. Thus at the beginning of phase 2 we have $\rho = r < n-2$. Now we compare $y_1 : a$. If they are of the same color, we throw y_1 and y_2 away. This leaves δ and ρ unchanged. In case they have different colors, we throw y_1 and a ball of R away (if $R \neq \emptyset$). This gives $\delta \to \delta$, and $\rho \to \rho - 1$ (or $\rho \to \rho$). Next take y_j with the smallest index and compare it to a, proceeding in the same fashion. Finally, if there is a top element y in the list left, we compare it to a. If they have the same color, put y into the reservoir, otherwise throw y and a ball of R away (if $R \neq \emptyset$).

We make the following observations on ρ.

e) If ever $\rho \le d-2$, then $\delta \le \rho+1 \le d-1$, and there is no k-majority. In particular, this holds when $r \le d-2$ at the start of phase 2.

f) Suppose then that we always have $\rho \ge d-1$. If $\rho \ge d$, then the balls in the reservoir are k-majority balls. On the other hand, if $\rho = d-1$, then there is no k-majority, since the remaining balls a, b in L have different colors.

It remains to estimate the number of tests in phase 2. Let t be the number of those tests with answer "no". Set $r = d-1+s$, where $s \ge 0$ and $t \le s$, since $\rho \ge d-1$. For the total number m of tests in phase 2 we obtain therefore

$$m \le t + \frac{n-2-r-t+1}{2}$$

where -2 accounts for the balls a and b, and $+1$ for the possible first ball. Hence

$$m \le \frac{n-r+t-1}{2} = \frac{n-d-s+t}{2} \le \frac{n-d}{2} = n-k,$$

and so

$$L \le (n-1)+m \le 2n-k-1. \qquad \blacksquare$$

Let us briefly discuss so-called *competitive algorithms* (see [7, Ch. 7]). We are again given the k-majority problem, with $2 \le c \le n$. Paul uses algorithms which finish the game after at most $2n-k-1$ tests (according to Theorem 3), but Carole is only allowed to use c colors (unknown to Paul). Let us denote the corresponding length by $M_c(n,k)$. Clearly,

(13) $$M_2(n,k) \le M_3(n,k) \le \cdots \le M_n(n,k) = 2n-k-1,$$

and

(14) $$M_c(n,k) \ge L_c(n,k) \quad \text{for all} \quad n,k,c.$$

The following theorem (which is difficult) solves the competitive problem apart from small c.

Theorem 4. *We have*

$$M_5(n,k) = 2n-k-1,$$

and hence

$$M_5(n,k) = M_6(n,k) = \cdots = M_n(n,k) = 2n-k-1.$$

In the light of Theorem 4 we make the following conjectures which would solve the whole problem. They are probably quite difficult and will require some new ideas.

Conjecture 2. *We have*

i) $L_3(n,k) = 2n - k - 1$

and thus

$$L_3(n,k) = L_4(n,k) = \cdots = L_n(n,k) = 2n - k - 1.$$

ii) $M_2(n,k) = 2n - k - 1,$

and thus

$$M_2(n,k) = M_3(n,k) = \cdots = M_n(n,k) = 2n - k - 1.$$

4. Third generalization: the plurality problem

We are given n balls and c colors as before with the same tests. Now Paul has to produce a ball z of *plurality color* (that is, the number of balls colored like z exceeds all other colors), or state that there is no plurality. Let us denote by $P_c(n)$ the number of tests required.

We clearly have $P_2(n) = L(n)$ and

$$P_2(n) \le P_3(n) \le \cdots \le P_n = \binom{n}{2}. \tag{15}$$

To see the last equality in (16), just note that if Carole always answers "no", then Paul has to make all $\binom{n}{2}$ comparisons.

Theorem 5. *We have*

$$P_3(n) \le \left\lfloor \frac{5n}{3} \right\rfloor - 2 \qquad (n \ge 2).$$

Proof. This is clear for $n \le 3$, so let us assume $n \ge 4$. We order the balls $x_1, x_2, \ldots, x_n$ and compare them one by one. The following set-up is useful.

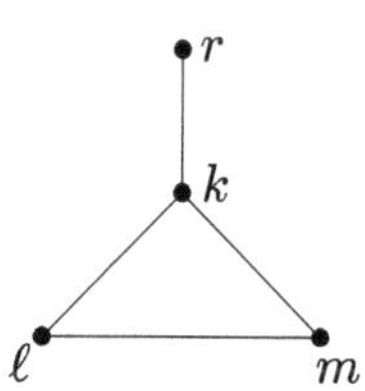

The figure means that there are k balls of color 1, ℓ balls of color 2 and m balls of color 3, whereas r balls may have color 2 or 3, but not color 1. In other words, an edge represents different colors. For example, after the first comparison we obtain the situation

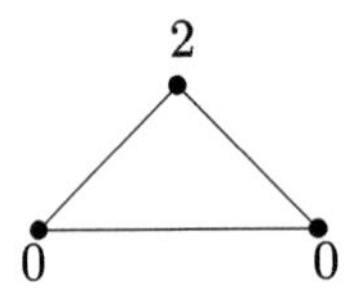

or

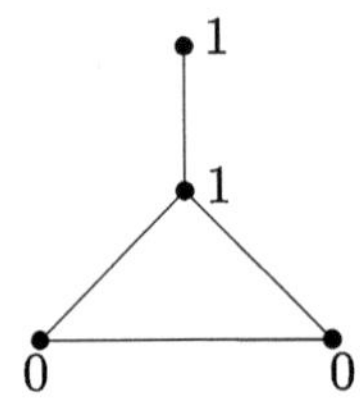

By comparing a new ball with an appropriate color-class the following lemma can be proved:

Lemma. *Any stage can be described by*

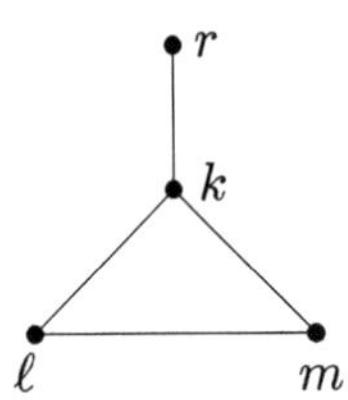

$r \geq 0$, and $k > \ell$ if $r > 0$

or

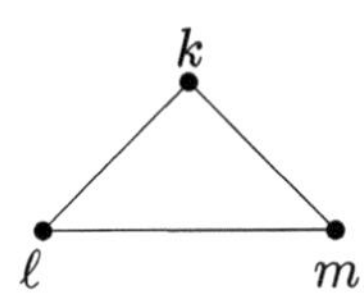

$k > \ell > m$.

Furthermore, if $a \geq b \geq c$ *are the color-numbers at this stage, then Paul has used at most* $2a + b + 2c - 2$ *tests so far.*

Let $a \geq b \geq c$ be the sizes of the color-classes, and $s = n - (a + b + c)$ be the number of remaining balls. By the lemma, Paul has used so far T tests with

$$T \leq 2a + b + 2c - 2.$$

Paul stops, when he encounters *for the first time* one of the following situations:

(I) $a = b = c$

(II) $a = b + s$

(III) $a = b + s + 1$.

Suppose (I) occurs. We test the remaining balls and obtain by induction for the number P of tests

$$P \leq T + \left\lfloor \frac{5(n-3a)}{3} \right\rfloor - 2 \leq 5a - 2 + \left\lfloor \frac{5n}{3} \right\rfloor - 5a - 2 = \left\lfloor \frac{5n}{3} \right\rfloor - 4.$$

Suppose (II) occurs. Then any ball of the a-group is a plurality ball, unless all remaining balls have the same color (different from a). We have

(16) $$a = b + s = n - a - c, \quad \text{hence} \quad n = 2a + c,$$

and thus

(17) $$T \leq 2a + b + 2c - 2 = 2n - 2a - 2c + b + 2c - 2 = 2n - 2a + b - 2.$$

Case i) $s = 0$. Then there is no plurality, and

$$L = T \leq 2n - a - 2 \leq 2n - \frac{n}{3} - 2 = \frac{5n}{3} - 2,$$

since $a \geq \frac{n}{3}$.

Case ii) $s \geq 1$. Take one of the remaining s balls, say x, and compare it to all the other $s-1$. If there is ever an answer "no", we are finished. Suppose they all have the same color (after $s-1$ tests). If $b = c$, we test $a : s$. If the answer is "yes", then a is the plurality group, if it is "no", then there is no plurality. Finally, if $b > c$, test $b : s$. If "yes", then there is no plurality, if "no", then a is the plurality group.

Altogether we obtain for the number P of tests

$$P \leq T + s \leq 2n - 2a + b - 2 + s = 2n - a - 2.$$

Since $n = 2a + c$ by (17), we have $a \geq \frac{n}{3}$, and hence again

$$P \leq \frac{5n}{3} - 2.$$

The case, when (III) occurs first, is dealt with similarly. ■

As for lower bounds it was shown in [3] that $P_3(n) \geq 3\lfloor \frac{n}{2} \rfloor - 2$. Both bounds are the best so far, but it is believed that the upper bound is the correct answer.

Conjecture 3. *We have*

$$P_3(n) = \left\lfloor \frac{5n}{3} \right\rfloor - 2 \quad (n \geq 2).$$

For $c \geq 4$ colors it is known that $P_c(n) = \Theta(cn)$, see [3]. Furthermore, there are some tight results for the probabilistic setting, where Paul uses randomized strategies [8, 10].

5. The liar problem

A closely related interesting problem is the following. Suppose in a room there are n people some of whom always tell the truth whereas others are unreliable (they sometimes speak the truth and sometimes they lie). The people in the room know about each other's reliability status. Now Paul enters the room and is required to determine the status of each person by asking questions of the form: He asks person x whether another person y is reliable. How many questions does he need in the worst case?

It is easy to see that, if there are at least as many liars in the room as reliable people, then Paul stands no chance to find out. Hence we assume that Paul knows that the number of reliable people is at least k with $k > \frac{n}{2}$. Let us denote the number of questions needed by $Q(n,k)$. The original problem (with ordinary majority $k = \lfloor \frac{n}{2} \rfloor + 1$) was solved by Blecher [6]:

$$Q\left(n, \left\lfloor \frac{n}{2} \right\rfloor + 1\right) = \left\lceil \frac{3n}{2} \right\rceil - 2.$$

In the same paper Blecher also noted the following general result:

Theorem 6. *We have*

$$Q(n,k) = 2n - k - 1.$$

We just sketch the proof of the upper bound, the lower bound is shown by a (quite involved) argument similar to the one used in [6]. Paul asks $x_1, x_2, \ldots$ about the reliability of a fixed person z. He stops when for the first time the number ℓ of "no" exceeds the number $\ell - 1$ of "yes", or when Carole has given the answer yes for the $(n-k)$-th time with, say, i answers "no".

In the first case, there are at least ℓ liars among $\{x_1, \ldots, x_{2\ell-1}\} \cup \{z\}$, hence in the remaining set of $n - 2\ell$ people there are at least $k - \ell$ reliable people, with $2(k - \ell) > n - 2\ell$. Now Paul determines the full partition in the remaining set, picks a reliable person w, and can now determine the status of z and the others with at most $\ell + 1$ further questions. Hence by induction the length L is bounded by

$$\begin{aligned} L &\leq (2\ell - 1) + Q(n - 2\ell, k - \ell) + \ell + 1 \\ &\leq 3\ell + 2n - 4\ell - k + \ell - 1 = 2n - k - 1. \end{aligned}$$

The second possibility is similarly dealt with. ■

Finally, there is the variant where Paul needs only find *one* reliable person, where he knows as before that there are at least k reliable people in the room, $k > \frac{n}{2}$. Let us denote by $R(n, k)$ the length of the liar game for this variant.

Before we present the result, we make some observations. If x is asked about the reliability of y, then we draw an arrow $x \overset{Y}{\to} y$, $x \overset{N}{\to} y$, depending on the outcome "yes" or "no". We write xT resp. xL if x is reliable resp. unreliable. Now

$x \overset{Y}{\to} y$ is compatible with xT, yT; xL, yT; xL, yL but not with xT, yL.

(18) Hence if $x \overset{Y}{\to} y$ and yL, then xL must hold.

Similarly,

$x \overset{N}{\to} y$ is compatible with xT, yL; xL, yT; xL, yL but not with xT, yT.

(19) Hence $x \overset{N}{\to} y$ implies that at least one of x or y is a liar.

Theorem 7. *We have*

$$R(n, k) = 2(n - k) - B(n - k).$$

Proof. Again we restrict ourselves to the upper bound. For the lower bound see [2]. Paul picks $2(n - k) + 1$ people, and *knows* that the reliable people are in the majority. So it suffices to prove the inequality for the ordinary majority liar game:

$$R\left(n, \frac{n+1}{2}\right) \leq n - B(n), \quad n \text{ odd}.$$

We set up a data structure similar to the coloring problem. There are buckets $B_1, \ldots, B_s$ and a dump D.

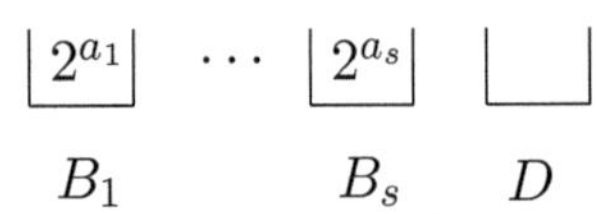

In every bucket B_i we have a "cube-like" structure with all answers "yes", and a unique sink z_i.

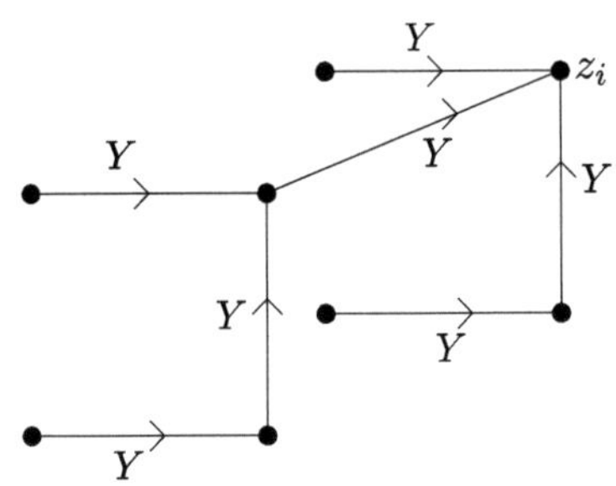

The initial configuration consists of n buckets each containing one person with D being empty. Note that we have used $2^{a_i} - 1$ questions to produce B_i. In the next step Paul picks two buckets B_i, B_j with $2^{a_i} = 2^{a_j}$, and asks the question $z_i \to z_j$.

If the answer is "yes", then he merges the buckets with z_j as new unique sink. On the other hand, if the answer is "no", then by (20) one of z_i or z_j must be a liar, say z_i. But then *all* people in B_i must be liars by (19). In this case we throw both buckets into the dump. It follows that D contains at any stage at least as many liars as reliable people.

The algorithm stops when the sizes of the buckets are all distinct, $2^{b_1} > 2^{b_2} > \cdots > 2^{b_t}$. The other possibility that all people are in the dump can clearly not occur. But then the unique sink z_1 of bucket B_1 must be reliable, since otherwise (observing (19)) the liars would outnumber the reliable people. Noting, as in the color game, that the number of questions is at most $n - B(n)$, the theorem follows. ■

6. A final generalization: determining the color classes

Let us consider a variant of the liar game which gives more information to Paul. Suppose he knows that as before the reliable people T always tell the truth, but that the unreliable people L *always* lie. Then if he asks x about the reliability of y and receives the answer "yes", then they are of the *same* type (T or L), while the answer "no" implies they are of *different* type. So in this restricted version the liar problem is just the ordinary color problem considered so far: "yes" corresponds to "same color" and "no" to "different color".

Another way to view this game is to consider bipartite graphs. We are given an (unknown) complete bipartite graph on n vertices. Whenever Paul tests two vertices $x : y$, he receives the answer "yes" (same color-class = no edge between x and y) or "no" (different color-class = there is an edge between x and y). If Paul has no previous information, then he needs $n-1$ questions, since there are 2^{n-1} possible bipartitions. We have already noted this in (2). Suppose now that Paul knows beforehand that the larger color-class has size at least $k > \frac{n}{2}$. Let us denote by $S(n,k)$ the length of the game in this case. Clearly,

$$(20)\qquad 0 = S(n,n) \le S(n,n-1) \le \cdots \le S\left(n, \left\lfloor \frac{n}{2} \right\rfloor + 1\right) = n-1,$$

where the last equality holds by the information-theoretic bound.

As an example, consider the first interesting case $S(n, n-1)$. We have

$$(21)\qquad S(n,n-1) = \left\lceil \frac{n+1}{2} \right\rceil \qquad (n \ge 3).$$

For the upper bound, Paul makes $\left\lceil \frac{n+1}{2} \right\rceil - 1$ disjoint comparisons, with one vertex left when n is odd. If there is a "no" in one of these comparisons, then one more comparison will determine the color-classes. And if the answer is always "yes", then Paul needs one more comparison to determine the status of the remaining vertex.

To see the lower bound, Carole answers "yes" to the first $\ell - 1$ comparisons, where $n = 2\ell$ or $n = 2\ell + 1$. Hence if n is even, there are at least two vertices whose status is undetermined, and if n is odd, there are three such vertices. Together with the possibility that all vertices have the same color, we find that two more tests are needed by the information-theoretic bound.

Two other small cases are known:

$$S(n, n-2) = \left\lceil \frac{2n+1}{3} \right\rceil \qquad (n \geq 5)$$

$$S(n, n-3) = \left\lceil \frac{7n+2}{10} \right\rceil \qquad (n \geq 7),$$

and further

$$S(n, n-4) = \frac{3}{4}n + O(1).$$

In general, this variant seems to amount to an intricate number-theoretic problem. The growth of $S(n, k)$, that is the coefficient of n, is not known for arbitrary k.

References

[1] M. Aigner, *Combinatorial Search,* Wiley (1988).

[2] M. Aigner, Variants of the majority problem, *Discrete Applied Math.,* **137** (2004), 3–25.

[3] M. Aigner, G. de Marco and M. Montangero, The plurality problem with three colors and more, to appear in *Theoretical Computer Science.*

[4] L. Alonso, E. Reingold and R. Schott, Determining the majority, *Information Processing Letters,* **47** (1993), 253–255.

[5] L. Alonso, E. Reingold and R. Schott, Average-case complexity of determining the majority, *SIAM J. Computing,* **26** (1997), 1–14.

[6] P. Blecher, On a logical problem, *Discrete Math.,* **43** (1983), 107–110.

[7] D. Du and F. Hwang, *Combinatorial Group Testing,* World Scientific (1993).

[8] Z. Dvořák, V. Jelínek, D. Král and J. Kynčl, *Probabilistic strategies for the partition and plurality problems,* manuscript.

[9] M. Fisher and S. Salzberg, Finding a majority among n votes, *J. Algorithms,* **3** (1982), 375–379.

[10] D. Král, J. Sgall and T. Tichý, *Randomized strategies for the plurality problem,* manuscript.

[11] M. Saks and M. Werman, On computing majority by comparisons, *Combinatorica,* **11(4)** (1991), 383–387.

[12] G. Wiener, Search for a majority element, *J. Statistical Planning and Inference,* **100** (2002), 313–318.

Martin Aigner

Freie Universität Berlin
Institute of Mathematics II

aigner@math.fu-berlin.de

BOLYAI SOCIETY
MATHEMATICAL STUDIES, 16

Entropy, Search, Complexity, pp. 27–70.

Coding with Feedback and Searching with Lies*

C. DEPPE

This paper gives a broad overview of the area of searching with errors and the related field of error-correcting coding. In the vast literature regarding this problem, many papers simultaneously deal with various sorts of restrictions on the searching protocol. We partition this survey into sections, choosing the most appropriate section for each topic.

1. Introduction

We consider the problem of transmitting messages over a noisy binary channel with noiseless feedback. This problem is closely connected to sequential binary search with errors. In [9] and [11] we can find many search problems which are equivalent to a coding problem. In chapter 8 of [39] one can find some results about binary search. Hill wrote a survey about searching with lies [42] in 1995. In this survey he explained very well the results with a fixed number of objects. In this survey we also report about the results which were in Hill's survey and additionally report about some new results and new models of the last five years. See [30] for a recent survey containing a discussion of the logical aspects of feedback search with lies. In 2002 Pelc wrote a survey "Searching games with errors – fifty years of coping with liars" ([68]). This survey is a non-technical survey. We will give a more technical survey with formal definitions and also some proofs. While our survey is partitioned into sections, choosing the most appropriate section

*Supported in part by INTAS-00-738.
A survey of results in coding with feedback and searching with lies until 2000.

for each topic was sometimes not an easy task. This is so because many papers simultaneously deal with various sorts of restrictions on the searching protocol. In each section we will make cross-references to the sections in which we report about papers which also could be in this section.

Rényi [69] reported the following story about the Jew Bar Kochba in 135 CE, who defended his fortress against the Romans.

> It is also said that Bar Kochba sent out a scout to the Roman camp who was captured and tortured, having his tongue cut out. He escaped from captivity and reported back to Bar Kochba, but being unable to talk, he could not tell in words what he had seen. Bar Kochba accordingly asked him questions which he could answer by nodding or shaking his head. Thus he acquired from his mute scout the information he needed to defend the fortress. It occurred to me that, if the story of Bar Kochba were true, then he would have been the forefather of information theory.

At the beginning of the 19th century the so-called Bar-Kochba game was very popular in Budapest. In this game, one player has to find out, by asking yes/no-questions, what the second player has in mind. In 1956 Shannon [73] introduced the discrete memoryless channel with noiseless feedback. He proved that the forward capacity is the same as without feedback, but the zero-error capacity is in some cases bigger with feedback than without. In 1961 Rényi [70] introduced the Bar-Kochba game with a given percentage of wrong answers. He described a sequential and a non-sequential version of the game in the introduction of the paper. He solved the non-sequential problem to find the minimal number of questions to determine the searched number with a certain probability, if the answers are correct with a given probability and the questions are chosen at random. He also remarked that the problem is connected with the coding problem in information theory. In 1964 Berlekamp considered in his dissertation the following coding scheme.

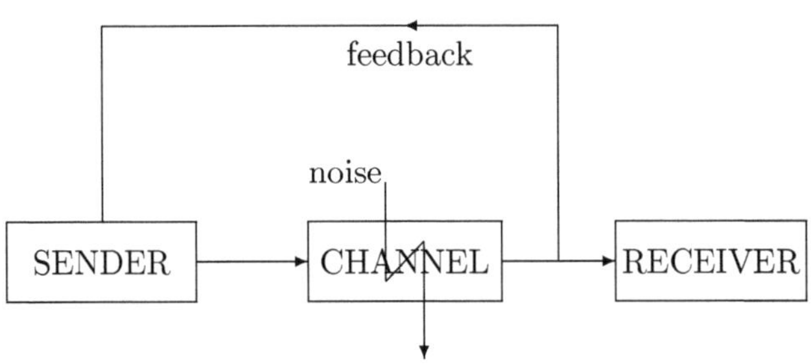

A sender wants to transmit a message $x \in \mathcal{X}$ over a noisy binary channel. $\mathcal{X} = \{1, \ldots, N\}$ denotes the set of possible messages and $Y = \{0, 1\}$ the binary coding alphabet. We have a passive feedback, that means that the sender always knows what has been received. The codewords are elements of Y^n and a codeword is in the form of: $c(x, y^{n-1}) = \big(c_1(x), \ldots, c_n(x, y^{n-1})\big)$, where $c_i : \mathcal{X} \times Y^{i-1} \to Y$ is a function for the i-th code letter which depends on the message we want to transmit and the $(i-1)$ bits which have been received before. We suppose that the noise does not change more than $l \in \mathbb{N}_0$ bits of a codeword. Berlekamp's idea was to consider each transmission as the following quiet-question-noisy-answer-game: The sender and the receiver have a common partition strategy. After the sender has chosen a message, the receiver chooses a subset S of the set of messages and asks if the message was among the subset S ($S \subset \mathcal{X}$). The sender sends "1" for yes and "0" for no over the noisy channel. Then the receiver chooses a new subset where his choice depends on the answer etc. The receiver tries to get the message with n questions and a jammer (the noise) wants to avoid this by changing at most l answers.

Later in 1976 Ulam [78] suggested independently an interesting two-person search game:

> Someone thinks of a number between one and one million. Another person is allowed to ask up to twenty questions, to each of which the first person is supposed to answer only yes or no. Now suppose one were allowed to lie once or twice, then how many questions would one need to get the right answer.

Obviously this binary sequential search problem with errors is equivalent to Berlekamp's quiet-question-noisy-answer-game and to the Bar-Kochba game with lies. Ulam raised this problem in 1976; that was twelve years after Berlekamp considered the block coding with feedback and fifteen years after Rényi's paper. At first the authors did not remember that the problem was much earlier known and have been considered by Berlekamp and Rényi. For this reasons it is called the Ulam–Rényi game. In 1992 Spencer [76] presented another aspect of Ulam–Rényi's game. He considered the following two person game. We take a board with two columns and $l + 1$ rows. The rows are numbered from l to 0 and the columns by two and one. A field with some chips on it corresponds to every row. Each round of the game is played in three steps. At the first step Paul distributes the chips of the field on the corresponding columns. At the second step Carole chooses one column. All chips in this column are shifted by one row down. The chips in row 0 and the selected column are removed. At step three all chips of one row are taken on its corresponding field. Then the round is finished. The game is terminated if every chip up to one is removed. The aim of Carole is to get the number of rounds as large as possible whereas Paul wants to get a small number of rounds. Also this game is equivalent to the Ulam–Rényi game. Throughout this paper we shall call Carole and Paul the two players. This idea goes back to Spencer, who also explained: Paul corresponds to Paul Erdös, who always asked questions and Carole corresponds to an ORACLE, whose answers need to be wisely evaluated.

2. Definitions and terminology

In each round of the game Paul gets a negative vote for a subset $T \subset \mathcal{X}$. If Paul gets more than l negative votes for a number, then this number cannot be the searched one because Carole is allowed to lie at most l times. Therefore, in each round we consider for all $0 \leq j \leq l$ the sets

$$S_j \triangleq \{x \in \mathcal{X} : \text{Paul got } (l-j) \text{ negative votes for } x\}.$$

Definition 2.1. The vector

$$\underline{v} = \big(|S_l|, |S_{l-1}|, \ldots, |S_0|\big) = (v_l, v_{l-1}, \ldots, v_0)$$

is referred to as a state (of the game). $\underline{v}$ is called a k-state if k questions are left.

Neither the states nor the dividing questions depend on a specific number, which is chosen by Carole. Everything depends only on the cardinality of the sets S_j.

Definition 2.2. Let $\underline{s}$ be an arbitrary state. The question if "$x \in S$" ($S \subset \mathcal{X}$) is introduced as a vector $[\underline{u}] = [u_l, \ldots, u_0]$, where $u_i \triangleq |S_i \cap S|$. The state $\underline{x}$ is reduced to the states $\underline{y}(\triangleq \text{YES}^{\underline{s}})$ and $\underline{z}(\triangleq \text{NO}^{\underline{s}})$ by the question $[\underline{u}]$ if there exists a $\underline{v} \geq 0$ such that

1. $\underline{x} = \underline{u} + \underline{v} \triangleq (u_l + v_l, u_{l-1} + v_{l-1}, \ldots, u_0 + v_0)$,
2. $y_i = u_i + v_{i+1}$,
3. $z_i = v_i + u_{i+1}$.

Definition 2.3.

1. Let $\underline{v}$ be an arbitrary state and let $\lfloor \underline{x} \rfloor$ be a question. The question is called legal if
$$0 \leq x_i \leq v_i, \quad \text{for all} \quad i = 0, \ldots, l.$$

Definition 2.4.

1. A 0-state $\underline{x}$ is called winning if $\sum_{i=0}^{l} x_i \leq 1$. Otherwise, $\underline{x}$ is called losing.
2. A k-state $\underline{x}$ is called winning if it can be reduced to two winning $(k-1)$-states. Otherwise, $\underline{x}$ is called losing.
3. A winning k-state is called borderline winning if it is a losing $(k-1)$-state.
4. A k-state is called singlet if $\sum_{i=0}^{k} |S_i| = 1$.
5. A k-state is called doublet if $\sum_{i=0}^{k} |S_i| = 2$.

Proposition 2.1.

1. *A winning n-state is also a winning k-state if $k > n$.*
2. *All borderline winning 1-states have the form $(0, \ldots, 0, 2)$.*
3. *Let $\underline{x}$ be a winning k-state and let $\underline{y}$ be some state such that $y_i \leq x_i$ holds for all $i \leq l$. Then $\underline{y}$ is also a winning k-state.*

Definition 2.5. For a given state $\underline{x}$, the function

$$V_n(\underline{x}) \triangleq \sum_{i=0}^{l} x_i \sum_{j=0}^{i} \binom{n}{j}$$

is called the n-th volume of $\underline{x}$.

The n-th volume of $\underline{x}$ can be interpreted as follows. We surround all messages in S_j (Carole can lie j times at all these numbers) by a sphere of radius j. Berlekamp used Pascal's Identity to show the following theorem.

Theorem 2.1 (Berlekamp's Conservation of Volume [18]). *Let $\underline{x}$ be a state which can be reduced to the states $\underline{y}$ and $\underline{z}$. Then*

$$V_n(\underline{x}) = V_{n-1}(\underline{y}) + V_{n-1}(\underline{z}).$$

With this theorem we get (by induction) thc following bound.

Theorem 2.2 (Berlekamp's Volume Bound [18]). *Let $\underline{x}$ be a winning n-state. Then*

$$V_n(\underline{x}) \leq 2^n.$$

We denote by $L_l(N)$ the minimal number of questions of Paul to find the searched number by using an optimal strategy.

Corollary 2.2 (The Hamming bound). *Let $|\mathcal{X}| = N$ and $l \in \mathbb{N}$. Then*

$$L_l(N) = n \Longrightarrow N \leq \frac{2^n}{\sum_{j=0}^{l} \binom{n}{j}}.$$

Definition 2.6. Let $\underline{s}$ be an arbitrary state. The number

$$ch(\underline{s}) \triangleq \min\left\{k \, : \, V_k(\underline{s}) \leq 2^k\right\}$$

is called the character of $\underline{s}$.

Obviously the character is a lower bound for the number of questions. We shorten the character of the starting state $(N, 0, \ldots, 0)$ with $ch_l(N)$, where l is the number of lies. In some papers the character is called the Berlekamp number.

Definition 2.7. A state $\underline{v}$ is called nice if it is a winning k-state and $k = ch(\underline{v})$.

3. A Berlekamp strategy and the translation bound

The following property of the states is proved by Berlekamp and called the Translation Bound.

Theorem 3.1 (Berlekamp [18]). *Let $\underline{x}$ be a state with $\sum_{i=0}^{k} x_i \geq 3$ and $n \geq 3$. If x is a winning n-state, then $T(x)$ is a winning $(n-3)$-state.*

From this theorem the following useful corollary follows.

Corollary 3.1.

1. $L_l(N) \leq k \Rightarrow L_{l-1}(N) \leq k - 3.$
2. $L_l(N) = k \Rightarrow L_{l+1}(N) \geq k + 3.$

We consider the following state-table, which was presented by Berlekamp:

Table 3.1.

column / row	1	2	3	4	5	6	7	8	9	10	11	12	13	14
1	4	2	1	1	1	1	1	1	1	1	1	1	1	1
2	8	6	4	1	0	0	0	0	0	0	0	0	0	0
3	36	22	14	10	5	1	0	0	0	0	0	0	0	0
4	152	94	58	36	24	15	6	1	0	0	0	0	0	0
5	644	398	246	152	94	60	39	21	7	1	0	0	0	0
6	2728	1686	1042	644	398	246	154	99	60	28	8	1	0	0
7	11556	7142	4414	2728	1686	1042	644	400	253	159	88	36	9	1
⋮	⋮	⋮	⋮	⋮	⋮	⋮	⋮	⋮	⋮	⋮	⋮	⋮	⋮	⋮

Definition 3.1.

1. This table is defined recursively. Let $A_{i,j}$ be the value in the i-th row and j-th column. The first two rows are the initial rows. They are given by:

$$A_{1,1} := 4, \quad A_{1,2} := 2, \quad A_{1,k} := 1 \ \forall\ k \geq 3$$

$$A_{2,1} := 8, \quad A_{2,2} := 6, \quad A_{2,3} := 4, \quad A_{2,4} := 1, \quad A_{2,k} := 0 \ \forall\ k \geq 5$$

The remainder of the table is derived recursively by: $\forall i \geq 3$ holds:

$$\text{with } j \geq 3 : A_{i,j} := A_{i-1,j-1} + A_{i-1,j-2} \tag{3.1}$$

$$\text{with } j = 2 : A_{i,2} := A_{i,3} + A_{i-1,1} \tag{3.2}$$

$$\text{with } j = 1 : A_{i,1} := A_{i,2} + A_{i,3} \tag{3.3}$$

2. Let $j, m \in \mathbb{N}$, then we set: $\underline{A}_{m,j} := \begin{pmatrix} A_{1,j} \\ A_{2,j} \\ \vdots \\ A_{m,j} \end{pmatrix}$.

Berlekamp's state table has the following properties.

Theorem 3.2 (Berlekamp [18]). *Let $A_{i,j}$ be the value in the i-th row and j-th column of Table 3.1.*

- *An $\underline{A}_{m,j}$-state is a winning $(3m - j)$-state for all $0 \leq j \leq 3m$.*
- *An $\underline{A}_{m,j}$-state can be reduced to $\underline{A}_{m,j+1}$ and $\underline{A}_{m-1,j-2}$.*
- *If $0 \leq j \leq 3 \leq i$, then holds: $A_{i,j} = 2\left(\frac{1+\sqrt{5}}{2}\right)^{3i-j-2} + 2\left(\frac{1-\sqrt{5}}{2}\right)^{3i-j-2}$.*

4. Linear growth

Berlekamp analyzed the asymptotic relationship between the error-correcting fraction $f = \frac{l}{n}$ and the rate $R = \frac{\log|\mathcal{X}|}{n}$. He proved an upper asymptotic bound. With the Translation Bound (Theorem 3.0) and the Hamming Bound 2.2 we can prove the following theorem.

Theorem 4.1 (Berlekamp's Tangential Bound). *Let f, R be like before, $l \in \mathbb{N}$ the maximal number of errors (lies), $k > 1$, $N = 2^k$ and $L_l(N) = n$, then holds:*

$$R \leq \begin{cases} 1 - h(f) + o(1), & \text{if } 0 \leq f \leq f_t, \\ -\dfrac{R_0}{f_0} f + R_0 + o(1), & \text{if } f_t \leq f \leq \frac{1}{3}. \end{cases}$$

He showed that this bound is attainable for small rates R $(0 \leq R \leq R_0 \approx 0.7)$ or big error-correcting fractions f $\left(0.2 \approx f_t \leq f \leq \frac{1}{3}\right)$. In [72] another way how to attain the bound for small rates is shown. Zigangirov [81] showed in 1976 that the Berlekamp bound is also attainable for big rates or small error-correcting fractions. He used a modified coding procedure which was developed by Horstein [44]. Finally we get the result:

Theorem 4.2.

Let p be selected in a way that $R = 1 - h(p)$. Then for $f_0 = \frac{1}{3}$, $R_0 = 0.6942$, $R_t = 0.29650$, $f_t = 0.19095$ and $\lim\limits_{n\to\infty} \frac{l(n)}{n} = f \leq \frac{1}{2}$ holds:

$$f = \begin{cases} f_0 - \dfrac{f_0}{R_0} R + o(1), & \text{if } 0 \leq R \leq R_t, \\ p + o(1), & \text{if } R_t \leq R. \end{cases}$$

5. Fixed error

In this section we will consider the function $L_l(N)$ if $l \in \{1, 2, 3\}$. Therefore we need some definitions.

Definition 5.1. Let $\underline{v}$ be a state with $ch(\underline{v}) = k$. Let $[\underline{x}]$ be a question, which reduce $\underline{v}$ to the states $\underline{s}$ and $\underline{t}$.

1. We call $[\underline{x}]$ balanced, if $ch(\underline{s}) = ch(\underline{t}) = k - 1$.
2. We call $[\underline{x}]$ a splitting, if $x_i = \lceil \frac{v_i}{2} \rceil$ for $0 \le i \le l$.

In 1987 Pelc [66] gave a general solution for the number of answers if $l = 1$. He developed an optimal strategy. The best first question is a splitting. Then he showed that there exist balancing questions for the remaining ones and he also describes them. Thus the character of the state after the first question gives the solution of the problem.

Theorem 5.1 (Pelc). *Let $\underline{s} = (N, 0)$ and $k = ch(\underline{s})$.*

1. *Let $N \in \mathbb{N}$ be even. Paul wins the game with k questions.*
2. *Let $N \in \mathbb{N}$ be odd.*
   (a) *Paul wins the game with k questions, if*
   $$V_{k-1}\left(\left(\frac{N+1}{2}, \frac{N-1}{2}\right)\right) \le 2^{k-1}$$
   (b) *Paul wins the game with $k+1$ questions, otherwise.*

In 1988 Czyzowicz, Mundici and Pelc [32] showed

Theorem 5.2. *Let $N = 2^m$, $m \ge 3$, $\underline{s} = (N, 0, 0)$ and $k = ch(\underline{s})$. Then Paul needs k questions to win the game.*

In 1990 Guzicki [40] solved the problem for $l = 2$. He also gave an optimal strategy, if $N \ge 90$ and calculated all remaining values. The first question in this strategy is also a splitting. The second question depends on the value of $N \bmod 4$ and $ch_l(N) \bmod 4$. The other questions are balanced. Thus the character of the state after the first two questions gives the solution of the problem, if $N \ge 90$.

Theorem 5.3 (Guzicki). *Let $\underline{s} = (N, 0, 0, 0)$ and $k = ch(\underline{s})$.*

Case $N < 90$:

1. *Paul needs $k+1$ questions to win the game if $N \in M_1 = \{3, 4, 5, 6, 9, 10, 11, 17, 18, 29, 30, 51, 89\}$*
2. *Paul needs k questions to win the game if $N \notin M_1$.*

Case $N \ge 90$:

1. *Paul needs k questions to win the game if*
   (a) *$N = 4z$ and $N\frac{k^2+5k+9}{2} \le 2^{k+2}$.*
   (b) *$N = 4z + 1$ and*

i. $k = 4a$ and $z\frac{k^2+k+2}{2} + (2z + a + 1)(k + 1) + n - a \leq 2^k$.

ii. or $k = 4a + 1$ and $(2z + 1)\frac{k^2+3k+4}{2} + 2z(k + 2) \leq 2^{k+1}$.

iii. or $k = 4a + 2$ and $(z + 1)\frac{k^2+k+2}{2} + (2z - a)(k + 1) + n + a \leq 2^k$.

iv. or $k = 4a+3$ and $z\frac{k^2+k+2}{2} + (2z+a+2)(k+1)+n-a-1 \leq 2^k$.

(c) $N = 4z + 2$

i. $k = 4a$ and $(z + 1)\frac{k^2+k+2}{2} + (2z + a + 2)(k + 1) + n - a \leq 2^k$.

ii. or $k = 4a + 1$ and $z\frac{k^2+k+2}{2} + (2z + a + 2)(k + 1) + n - a \leq 2^k$.

iii. or $k = 4a + 2$ and $z\frac{k^2+k+2}{2} + (2z + a + 2)(k + 1) + n - a \leq 2^k$.

iv. or $k = 4a + 3$ and $N\frac{k^2+5k+9}{2} \leq 2^{k+2}$.

(d) $N = 4z + 3$ and

i. $k = 2a$ and $(2z + 2)\frac{k^2+3k+4}{2} + (2z + 1)(k + 2) \leq 2^{k+1}$.

ii. or $k = 2a + 1$ and $(2z + 2)\frac{k^2+3k+4}{2} + (2z + 1)(k + 2) \leq 2^{k+1}$.

1. *Paul needs $k + 1$ questions to win the game, otherwise.*

In 1992 Negro and Sereno gave a solution for $l = 3$, if $N = 2^m$. They consider the set $N = \{0, \ldots, 2^m - 1\}$ and the binary representation of the objects. The first optimal m questions are splittings. They showed that there exists remaining balanced questions, if $m > 3$.

In 1998 we gave a general solution for $l = 3$ [36]. Again we have the situation that we can give a general strategy if N is big enough (≥ 266). It turns out, that we can decide after the first three optimal questions, how many questions are necessary for Paul to get the correct answer. After the first three questions there exists a strategy that forms the next question until there are 15 questions left. Unfortunately, we cannot give a similar strategy for the last 15 questions. We use an algorithm for the last 15 questions. This algorithm was developed by Guzicki [40] to solve the Ulam–Rényi game in the case of two lies. It is easy to prove that we can also use this algorithm for three lies. The main theorem is as follows.

Theorem 5.4 (Deppe). *Let $\underline{s} = (N, 0, 0, 0)$ and $k = ch(\underline{s})$.*

Case $N < 266$:

1. *Paul needs $k + 2$ questions to win the game if $N \in M_2 = \{3, 5\}$.*
2. *Paul needs k questions to win the game if $N \in M_0 = \{1, 2, 14, \ldots, 16, 22, \ldots, 28, 35, \ldots, 50, 57, \ldots, 88, 95, \ldots, 154, 158, \ldots, 264\}$.*
3. *Paul needs $k + 1$ questions to win the game if $N \notin M_0 \cup M_2$.*

Case $N \geq 266$:

1. *Paul needs* k *questions to win the game if*

$$V_k(\underline{s}) + a_1(N)\binom{k-1}{3} + 2a_2(N,\underline{t})\binom{k-2}{2} + 4a_3(N,\underline{u})[k-3] \leq 2^k,$$

where $\underline{t}$ *is the state after the second question and* $\underline{s}$ *is the state after the third question of the mini-strategy.*

2. *Paul needs* $k+1$ *questions to win the game, otherwise.*

6. Asymptotic results for fixed error

In the previous section we have seen, that for $l \in \{1,2,3\}$ and N big enough the first l questions decide whether we can solve the game with the character of the initial state steps or we need one more question. In 1992 Spencer proved the following

Theorem 6.1 (Spencer [76]). *There exists constants* $c(l)$, $q_0(l)$ *such that for all states* $\underline{v}$ *with* $ch(\underline{v}) \geq q_0(l)$*, the following is true: If* $v_0 > c(l)$*, then* $\underline{v}$ *is a nice state.*

Using this statement Spencer proved:

Theorem 6.2 (Spencer [76]). *Let* $\underline{s} = (N, 0, \ldots, 0)$ *be the initial state,* $k = ch(\underline{s})$ *and let* $\underline{v}$ *be the state after the first* l *questions. Then there exists a constant* $q(l)$*, such that the following is true: If* $k > q(l)$*, then Paul wins the game starting from state* $\underline{v}$ *with* $ch(\underline{v})$ *questions.*

From this theorem we get the following.

Corollary 6.1.

- *If* $N = 2^k$*, then for all* l *there exists* $N_0 : \forall N \geq N_0 : L_l(N) = ch_l(N)$.
- *In general holds for all* l *there exists* $N_0 : \forall N \geq N_0 : L_l(N) = ch_l(N) + l$.

In [35] we improve Corollary 6.1. We show the following

Theorem 6.3. *For all* l *there exists* $N_0 : \forall N \geq N_0 : L_l(N) \leq ch_l(N) + 1$.

The proof of this theorem is a generalization of the theorem which is used for the case of three lies.

7. Fixed number of messages

For some special numbers $N_0 \in \mathbb{N}$ it is also possible to get an explicit formula $L_l(N_0) = f(l)$ by using Berlekamp's translation bound and his state table. We present some examples. The sets S_j were defined as follows:

$$S_j \triangleq \{x \in \mathcal{X} : \text{Paul gets } (l-j) \text{ negative votes for } x\}.$$

Obviously, this set is also defined for negative numbers. Therefore we can define a q-position state as follows:

Definition 7.1. Suppose that Paul has asked q questions. The vector

$$\underline{v}^q = \left(|S_l|, |S_{l-1}|, \ldots, |S_0|, \ldots, |S_{l-q}|\right) = (v_l, v_{l-1}, \ldots, v_0, \ldots, v_{l-q})$$

is referred to as a q-position state.

In the same way we define a q-position question.

Proposition 7.1.

1. *$L_l(2) = 1 + 2l$ for all $l \geq 0$.*
2. *$L_l(4) = 7 + 3(l-2)$ for all $l \geq 2$.*
3. *$L_l(16) = 4 + 3l$ for all $l \geq 0$.*
4. *In the general case: If one finds a strategy that reduces $(N_0, 0, \ldots, 0)$ to an q-position state $\underline{v}$, which is less or equal $\underline{A}_{l,j_0}$ taken from Berlekamp's state table and this strategy is optimal for some $l = c_0$, then for all $l \geq c_0$*

$$L_l(N_0) = f(l) = L_{c_0}(N_0) + 3(l - c_0).$$

Proof. The statements can be proved by using the first column of Berlekamp's state Table 3.1, Corollary 3.1 and the Hamming Bound. Further, assume that we can find a strategy (with q-position questions) which reduces the state in the worst case to an q-position state. If this vector is less than or equal to a column of the state table, then we can continue with the state table strategy. If we know that this strategy is optimal, then we can use the same strategy to solve the problem for one more lie with three additional questions. By using Corollary 3.1, we know that this is also optimal. ∎

The idea using a state table to solve the Ulam–Rényi game belongs to Hill and Karim [43]. The authors used another state table as compared to Table 3.1. They solved Ulam's problem for $N = 1000000$ and $l = 1, 2, 3$ by using this table. The table used does not harmonize with the translation bound of Berlekamp. Later they used Table 3.1 and get together with Berlekamp many theorems like the following.

Theorem 7.1. *For each $l \geq 8$ we have the identity*

$$L_l(1000000) = 3l + 26.$$

Similar statements can be found in the PhD-thesis of Karim [45] and DesJardins [37].

8. Fast algorithm

In the papers which we cited in the sections about fixed number of lies and about fixed number of objects, you also find an algorithm for the special cases ($l = 1, 2, 3$ and some values of N). In this section we want to present some general near-optimal algorithms. In most cases one of the ideas is to make the algorithm locally optimal. This means to reduce the actual game state to two states with equal volume. The reason for this idea is Berlekamp's Volume Bound (Theorem 2.1) and the Conservation of Volume (Theorem 2.1). An algorithm is optimal if the character of the state is reduced by one in every step. A very simple algorithm is to use splittings (Definition 5.1) as questions in every step. Splitting are locally optimal, if all entries of the state are even. But we get a better algorithm if we minimize the volume difference $\left| V_{k_1}(\text{YES}) - V_{k_1}(\text{NO}) \right|$ in every step. Such an algorithm is considered in [54]. In [23] the authors improve the bounds of [54] by introducing a novel Volume-minimization rule that also incorporates Berlekamp's Translation Bound. Another way to get a fast algorithm is shown in [15]. The author introduces a tree structure in such a way that the nodes contain integer-valued vectors (game-states). The vector contained in each node is uniquely determined by a special arithmetic operation applied to the vectors contained in its sons. More precisely, we notice that the identity $\sum_{i=0}^{e} \binom{n-l-i-1}{i} 2^{l-i} = \sum_{i=0}^{l} \binom{n}{i}$, can be used in the following way. We construct a regular rooted binary tree having $k^*_{l-i} = \binom{n-l-i-1}{i}$ terminal nodes at level i and associate a vector of length $l+1$ having 1 in the i-th position and 0's in all other positions with each of them, where $i = 0, \dots, l$. If two nodes located at some level t of the tree contain the vectors $a = (a_l, \dots, a_0)$ and $b = (b_l, \dots, b_0)$ and these nodes have a common parent at level $t+1$ then we specify that the parent must contain the vector $c = (c_l, \dots, c_l)$ having components $c_l = a_l + b_l$ and $c_j = a_j + b_j - c_{j+1}$, $j = 0, \dots, l+1$. It can be easily checked that if the tree is constructed in accordance with these rules and has the distribution of terminal nodes $(k_l, \dots, k_0)$ then the root contains the vector having 1 in the l-th position and 0's in all other positions. Now let $\underline{v}$ be any state vector of the Ulam–Rényi game. If we now construct a tree by the arithmetic operation of [15] with this state contained in the root, we get a strategy for the game starting at this vector. This strategy can be calculated very quickly. It turns out that the strategy is optimal for the case $l = 1$. The author gets also the following bound.

Theorem 8.1. *The strategy uses at most $ch_l(N) + 3l$ bits for N numbers.*

The problem of the computational complexity of coding and decoding algorithms in the Ulam–Rényi game is also important. Indeed the notion of "complexity" should not only be restricted to the number of questions or to the amount of feedback, but also should be concerned with the computational resources needed to formulate questions, and to deduce Paul's current state of knowledge as the result of Carole's answers. In [58] polynomial time complexity results are proved.

9. Comparison and interval questions

In 1997 Mundici and Trombetta ([59]) used bicomparison questions, asking "is x in $[a, b]$ or in $[c, d]$". They get the following result.

Theorem 9.1. *Let $l = 2$, $N = 2^m$ and $m > 2$. If Paul is restricted to ask bicomparison questions. Then he needs $ch\big((N, 0, 0)\big)$ questions.*

Aigner considered another special version of the Ulam problem [12]. In his version Paul is only allowed to ask comparison questions in form of: "Is the number less than or equal to i?" We denote Carole's guess number with x^*. Aigner analyzed the problem by refining the Berlekamp volume.

Definition 9.1. Let $x \in S_j \subset \mathcal{X}$ and $\underline{x}$ be a n-state, then we call:

$$v_n(x) \triangleq \sum_{i=0}^{j} \binom{n}{i}$$

the weight of x.

We denote by $L_l^{\leq}(N)$ the minimal number of questions of Paul to find the searched number with comparison questions by using an optimal strategy.

Remark 9.1. Let $\underline{x}$ be a state, then holds: $V_n(\underline{x}) = V_n \triangleq \sum_{x \in \mathcal{X}} v_n(x)$.

Definition 9.2. Let $\underline{x}$ be a n-state, $x \in \mathcal{X}$ and Paul's next question will be: "Is x^* less than or equal a?", then we set:

$$v_{n-1}^{1,a}(x) = \text{the new weight of } x, \text{ if the answer is yes}$$

$$v_{n-1}^{0,a}(x) = \text{the new weight of } x, \text{ if the answer is no}$$

$$V_{n-1}^{1,a} \triangleq \sum_{x \in \mathcal{X}} v_{n-1}^{1,a}(x)$$

$$V_{n-1}^{0,a} \triangleq \sum_{x \in \mathcal{X}} v_{n-1}^{0,a}(x)$$

Corollary 9.2 (conservation of volume). *Let $a \in \mathcal{X}$, then holds:*

$$V_n = V_{n-1}^{1,a} + V_{n-1}^{0,a}$$

Aigner used the same arguments as Berlekamp to achieve the volume bound and the Hamming bound. The Hamming bound is a lower bound for the number of questions that Paul has to ask in order to guess x^*. Aigner achieves an upper bound by deriving a weak version of the volume bound.

Lemma 9.3. $\sum_{j=1}^{p} \binom{j}{l} 2^{-j} = 2 - 2^{-p} \sum_{j=0}^{l} \binom{p+1}{j} \ \forall p, l \geq 1$

With this lemma we get the following theorem.

Theorem 9.2. *Let $l \in \{1, 2\}$ be the number of lies. If $V_n(\underline{x}) \leq 2^{n-2l+1}$, then $\underline{x}$ is a winning n-state.*

With this theorem we get the following result.

Corollary 9.4. *Let* $|\mathcal{X}| = N$, *then holds:*

1. *For* $l = 1 : N \leq \frac{2^{n-1}}{1+n} \Longrightarrow L_1(N) \leq n$
2. *For* $l = 2 : N \leq \frac{2^{n-3}}{1+n+\binom{n}{2}} \Longrightarrow L_2(N) \leq n$

Spencer [75] showed the following theorem.

Theorem 9.3. $ch\big((N,0)\big) \leq L_1^{\leq}(N) \leq ch\big((N,0)\big) + 1$

A general bound of the number of comparison questions is given in [47]. They showed the following.

Theorem 9.4. *Let* k *be the character of the initial state and* $c = \min\Big\{n : 2^{n-l} \geq N \sum_{j=0}^{l} \binom{n-l}{j}\Big\}$. *Then holds*

$$k \leq L_l^{\leq}(N) \leq c$$

Spencer and Winkler [77] and Aslam and Dhagat [14] considered linearly bounded comparison questions; the result can be found in section 12.

In [77] the authors also considered the case with asymmetric comparison questions. The result can be found in the next section.

10. Asymmetric questions

In this section we will consider the half-lie variant of the Ulam–Rényi game. That means we are asking for the minimum number of questions, if at most l of the negative answers are lies. This model makes also sense in the coding problem. Here we are restricted to errors if we receive a 0. This problem is considered in [28] for the case of $l = 1$ and $N = 2^m$. The authors obtain the following results.

Definition 10.1. We denote by $H_l(N)$ the minimal number of questions of Paul to find the searched number by using an optimal half-lie strategy.

Theorem 10.1. *Let* $m \in \mathbb{N}$, $N = 2^m$, $C = \Big\{m \in \mathbb{N} : 2^m ch\big((N,0)\big) = 2^{ch((N,0))} - 1\Big\}$ *and* $E_s = \{m \in \mathbb{N} : m_s + 1 \leq m \leq m_s + s - 2\}$, *where* $m_s = \max\{m : m + s + 1 \leq 2^s\}$.

1. $ch\big((N,0)\big) - 2 \leq H_1(N) \leq ch\big((N,0)\big) - 1$.

2. $H_1(2^m) = ch\big((N,0)\big) - 1$ *for all* $m \in C$.
3. $H_1(2^m) = ch\big((N,0)\big) - 2$ *for each* $s \in \mathbb{N}$ *and all* $m \in E_s$.

The case $l = 2$ is considered in [31]. Like in the case $l = 1$ the number of needed questions is below the sphere packing bound.

Theorem 10.2. *There exists* $M_0 : \forall M \geq M_0\ H(M,2) \geq ch_2(M) - 4$.

Definition 10.2. $m_s = \max\left\{m : \binom{m+s}{2} + m + s \leq 2^s\right\}$

$m_{s,e}$ is the largest integer m such that $m + s$ questions are necessary to find an element over a search space with 2^m elements and e lies.

Theorem 10.3. *For all* $13 \leq s \in \mathbb{N}$ *and* $m = m_{s,2} + 1$ *it holds* $H_2(2^m) \leq ch_2(2^m) - 3$.

The authors of [31] also give an upper bound for the general case.

Theorem 10.4. *Let* $e \in \mathbb{N}$ *then there exists* $s_0 : \forall s \geq s_0\ H_e(2^m) \leq ch_e(2^m) - e$, *for each* m *satisfying* $m_{s,e} + 1 \leq m \leq \left\lceil \frac{m_{s+1,e}+m_{s,e}}{2} \right\rceil$.

For the general case another result can be found in [47]. There the game with asymmetric comparison questions is considered.

Theorem 10.5.

$$H_l^{\leq}(N) \geq \log N + l \log\log N + O(l \log l).$$

11. Detecting errors

In this section we will consider some modified rules of the game. Paul wins the game if he gets the unknown number or if he detects at least one lie. This version of the game was considered in [64]. We call this game interactive detecting game D. The author also defined a second version of the game. In this version Paul wins the game if he get the unknown number or if he detect at least one lie and can tell how many times Carole lied. This game is called interactive detecting game D^*.

Definition 11.1.

1. We define by $D(N,l)$ the minimal length of Paul's winning strategy in game D with N numbers and l lies.
2. We define by $D^*(N,l)$ the minimal length of Paul's winning strategy in game D^* with N numbers and l lies.

The following results are proved in [64].

Theorem 11.1.

$$D(N,l) = \lceil \log N \rceil + l$$

for all $l, N \in \mathbb{N}$.

Theorem 11.2.

1. $$D^*(N,2) = \lceil \log N \rceil + 3$$

   for all $N \in \mathbb{N}$.

2. $$\lceil \log N \rceil + l \leq D^*(N,l) \leq \lceil \log N \rceil + 3l$$

12. Linearly bounded error

Spencer and Winkler considered in [77] another restriction of the game. The same approach is done by Aslam and Dhagat in [14]. In their version Carole is permitted only to lie up to a fixed fraction r of the number of questions. This variant of the problem was originally considered in [63] for the the non-adaptive and the fully adaptive version of the problem. They considered three different models. In version A Carole has to answer in such a way that at no step the number of lies divided by the number of questions is greater than r. Now let q be the number of questions Paul has asked. Then in version B Carole has to answer in such a way that at no step the number of lies divided by the number of questions is greater than $r * q$. They also considered version C which is the non-adaptive version of version B. Spencer and Winkler get the following result and call it the three thresholds theorem.

Theorem 12.1.

1. *In version A Paul wins with $O(\log N)$ questions if $r < \frac{1}{2}$ but Carole wins if $r \geq \frac{1}{2}$.*
2. *In version B Paul wins with $O(\log N)$ questions if $r < \frac{1}{3}$ but Carole wins if $r \geq \frac{1}{3}$.*
3. *In version C Paul wins with $O(\log N)$ questions if $r < \frac{1}{4}$, Carole wins if $r > \frac{1}{4}$ and Paul wins with $O(N)$ questions, if $r = \frac{1}{4}$.*

The first item is proved independently by Aslam and Dhagat. They also looked at the unbounded case of version A. That means Carole chooses any positive integer. They get the following result.

Theorem 12.2. *In the unbounded case of version A Paul wins with $O(\log m)$ questions if $r < \frac{1}{2}$ and Carole thought of the number m.*

They also consider the version A with comparison questions.

Theorem 12.3. *In version A with comparison questions Paul wins with $o(n)$ questions if $r < \frac{1}{2}$.*

In [38] the authors also consider these models with extra restrictions in the questions.

Definition 12.1. If Paul is allowed only to ask questions like "Is the i-th bit in the binary representation of x equal to 1", we say he is only allowed to ask bit questions.

Theorem 12.4. *Let Paul and Carole play version B of the game.*

1. *Paul has a winning strategy with n bit questions if and only if $N \leq 2^{\left\lfloor \frac{n-\lceil rn\rceil}{\lceil rn\rceil+1} \right\rfloor}$.*
2. *Paul needs $n = \left\lceil \frac{8\log N}{(1-3r)^2} \right\rceil$ for a winning strategy with comparison questions if $r < \frac{1}{3}$.*

Theorem 12.5. *Let Paul and Carole play version C of the game.*

1. *Paul has a winning strategy with n bit questions if and only if $N \leq 2^{\left\lfloor \frac{n}{2\lceil rn\rceil+1} \right\rfloor}$.*
2. *Paul has a winning strategy with n comparison questions if and only if $N \leq \frac{n}{2\lceil rn\rceil+1} + 1$.*

13. Local restrictions

In this section, constraints on lie patterns are of local nature. We will present the results of [34].

Definition 13.1. We call $e^n = (e_1, \ldots, e_n)$ with

$$e_j = \begin{cases} 0, & \text{if Carole's } j\text{-th answer was wrong,} \\ 1, & \text{if Carole's } j\text{-th answer was correct,} \end{cases}$$

the error pattern of the game.

One possible local restriction is "Carole cannot lie twice successively". This means that in the error pattern e^n two consecutive ones are forbidden. We say that 11 is a forbidden lie pattern for Carole and denote it by $(1,1) \not\subset e^n$. The authors of [34] studied the feasibility of search with a forbidden lie pattern. They get the following result.

Theorem 13.1. *Search with the forbidden lie pattern p is feasible if and only if $p \in \{(0), (1), (0,1), (1,0)\}$.*

They also showed optimal algorithms for the four cases. For (0) and (1) it is trivial. For the forbidden lie patterns $(0,1)$ and $(1,0)$ they show a one-to-one correspondence between this game and the Ulam–Rényi game with one lie.

14. Unbounded search

In the case of unbounded search Carole thinks of an arbitrary integer x. Paul tries to guess this integer by asking for arbitrary subset of $\mathbb{N}$. For example "is the number even" or "is the number $\leq t$". The number of questions Paul needs to ask in order to get x is a function of x. The unbounded search problem without lies and with comparison questions was introduced by Bentley and Yao [17]. They proved the following theorem.

Theorem 14.1. *If $f(x)$ questions suffice to solve the unbounded search problem with comparison questions, then $f(x)$ satisfies Kraft's inequality:*

$$\sum_{x \geq 1} 2^{-f(x)} \leq 1.$$

They proved this theorem by realizing that the sequences of answers for each $x \in \mathbb{N}$ give a prefix code for the integers.

Definition 14.1.

1. Let $f : \mathbb{N} \to \mathbb{N}$ then $Kr(f) = \sum_{x \geq 1} 2^{-f(x)}$.
2. Let $x \in [0,1]$ be a real number, x is recursive if there is an algorithm that takes a rational number y as input and determines if $x < y$.

The following theorem of Beigel [16] show that Bentley and Yao's bound is essentially optimal.

Theorem 14.2. *Let f be a nondecreasing recursive function such that $Kr(f)$ is recursive. There is an algorithm that solves the unbounded searching problem by asking $f(x)$ questions if and only if $Kr(f) \leq 1$.*

Definition 14.2. A question is called irredundant if it cannot be deduced from previous answers.

Knuth [48] proved a theorem about unbounded search problem in which Paul is restricted to irredundant questions.

Theorem 14.3. *Let f be a monotone recursive function. There is an algorithm that Paul wins the unbounded searching problem by asking $f(x)$ irredundant questions if and only if $Kr(f) \leq 1$.*

Beigel also gives some bounds for the minimal number of questions ($f(x)$).

Definition 14.3.

$$\log^{(i)}(x) = \begin{cases} x & \text{if } i = 0, \\ \log\log^{(i-1)}(x) & \text{otherwise.} \end{cases}$$

Theorem 14.4.

1. *Paul can win the unbounded searching problem with comparison questions with $\left\lceil \sum_{i=1}^{\min\{t:\log^{(t)}(x)\leq 1\}} \log^{(i)}(x) \right\rceil + 2$ questions.*

2. *Let e be the Euler number,* $0 < \varepsilon < e$ *and*

$$f(x) = \left\lceil \left(\sum_{i=1}^{\min\{t:\log^{(t)}(x)\le 1\}} log^{(i)}(x) \right) \right.$$

$$\left. - \left(\log\log(e-\varepsilon) \right) \left(\min\left\{ t : \log^{(t)}(x) \le 1 \right\} \right) \right\rceil .$$

Paul can win the unbounded searching problem with comparison questions with $f(x) + O(1)$ *questions.*

There is also a further modification considered in [17]. They considered parallel questions.

Definition 14.4. We say that Paul asks p-parallel, if he ask p-questions in each round.

Beigel gives the following theorem for parallel questions.

Theorem 14.5. *Let* f *be a nondecreasing recursive function such that* $\sum_{n\geq 1}(p+1)^{-f(x)} \leq 1$ *and the value of the sum is recursive. Paul can win the unbounded searching problem with p-parallel comparison questions with* $f(x)$ *rounds of p questions.*

Spencer and Winkler [77] considered linearly bounded errors, the result can be found in section 12.

15. Optimal strategies with minimum adaptiveness

In many practical situations the goal is to find searching strategies with minimum adaptiveness. That means that many questions can be asked without feedback. Unfortunately, in general totally non-adaptive strategies are much worse (see for example [56]). The authors of the paper [25] found strategies with minimum adaptiveness which reach the sphere-packing bound. They consider message sets of cardinality 2^m. The first result is an existence theorem. They call a strategy optimal if the number of questions is equal to the character of the initial state.

Theorem 15.1.

1. *For every $l \in \mathbb{N}$ there exist an $m_0 \in \mathbb{N}$ such that there exists an optimal searching strategy in which questions can be submitted to Carole in only two rounds. In the first round we submit m questions and the remaining are submitted in the second round.*
2. *If $l = 3$ and $m \geq 99$ then there exists an optimal strategy in two rounds.*

The optimal strategy for the case $l = 3$ is explicitly given in [25]. The case $l = 1$ was settled by Pelc, who proved that adaptiveness has no role to decrease the number of questions. The case $l = 2$ was analyzed in detail in [28].

16. The continuous case

In [47] we find a generalized version of Berlekamp's volume bound. The authors consider a continuous search set instead of a finite one and they give an optimal strategy which achieves the generalized Volume bound. In this strategy they only use comparison questions.

In [20] the problem of identifying two distinguished elements (x, y) in the search space $S = \{(x, y) : x < y,\ x, y \in [0, 1[\}$ was considered. There were presented two non-sequential optimal algorithms that reach the general volume bound (Group Testing, Parity Testing).

We consider again a game between Paul and Carole. The difference to the first game is that Carole thinks of a number of a measurable set$(= S)$. Paul regards each answer as a vote against a certain subset of words. Paul asks n questions and Carole is allowed to lie at most l times.

Definition 16.1. If Paul has q questions left, in the situation described above for all $j \in \{1, \dots, l\}$ we define:

$$X_j^q \triangleq \{x \in S : x \text{ is the selected number, if } l - j \text{ questions were wrong}\}$$

Definition 16.2. We call the vector $\underline{v} = \big(\mu(X_l^q), \dots, \mu(X_0^q)\big) = (v_l, v_{l-1}, \dots, v_0)$ a q-state (of the game), where μ is the measure of X_i^q.

Definition 16.3. We say the state $\underline{x}$ can be reduced to the states $\underline{y}$ and $\underline{z}$, if $\exists \underline{u}, \underline{v} \geq 0$:

1. $\underline{x} = \underline{u} + \underline{v}$
2. $y_i = u_i + v_{i+1}$
3. $z_i = v_i + u_{i+1}$

Definition 16.4.

1. We call $e = (e_0, \ldots, e_{n-1})$ with

$$e_j = \begin{cases} 0, & \text{if Carole's } j+1\text{-th answer was wrong} \\ 1, & \text{if Carole's } j+1\text{-th answer was correct} \end{cases}$$

   a lying pattern of the game.
2. We set

$$E \triangleq \{e \,:\, e \text{ is a possible lying pattern of the game}\}$$

   the lying pattern set.

Remark 16.1. It holds: $|E| = \sum_{i=0}^{l} \binom{n}{i}$.

Definition 16.5. Let $\underline{x}$ be a q-state, then we call:

$$W_q(\underline{x}) \triangleq \sum_{i=0}^{l} \mu(X_i^q) \sum_{j=0}^{i} \binom{q}{j}$$

the weight of the q-state $\underline{x}$.

Remark 16.2. Because of the definitions holds: $W_n(\underline{x}) = \mu(S)|E|$.

Lemma 16.3 (conservation of weight). *Let $\underline{x}$ be a q-state which can be reduced to the $(q-1)$-states $\underline{y}$ and $\underline{z}$, then holds:*

$$W_q(\underline{x}) = W_{q-1}(\underline{y}) + W_{q-1}(\underline{z})$$

Lemma 16.4. *Let n be the number of questions, l be the number of lies, S be the search set and x be Carole's secret element, then a lower bound to the size of the set R in which Paul can confine the unknown pair (x, y) is given by*

$$\mu(R) \geq 2^{-(n+1)}|E|\mu(S)$$

In the group testing model Carole thinks of two numbers in $[0,1[$. Paul asks questions like: "Is at least one of the numbers in $R \subset [0,1[$?" The search problem can be regarded as a search in the search set $[0,1[\times [0,1[$. In this case there is no difference between (x,y) and (y,x), thus we can set $S = \{(x,y) : x,y \in [0,1[,\, x > y\}$. We number Paul's questions from 0 to $n-1$.

Definition 16.6.

1. We set $A_q \triangleq$ the subset Paul asks in his q-th question.
2. We set

$$C_q \triangleq \begin{cases} [0,1[\times A_q \cup A_q \times [0,1[, & \text{if Carole answers 1 to the } q\text{-th question} \\ \bar{A}_q \times \bar{A}_q, & \text{if Carole answers 0 to the } q\text{-th question} \end{cases}$$

Obviously $(x,y) \in C_q$, if Carole's answer was correct.

Definition 16.7. Let $E = \{e_1, e_2, \ldots, e_v\}$ be the lying pattern set, then we define:

$$C_q^j \triangleq \begin{cases} C_q, & \text{if the } q\text{-th entry of } e_j \text{ is } 0 \\ \bar{C}_q, & \text{if the } q\text{-th entry of } e_j \text{ is } 1 \end{cases}$$

Obviously $(x,y) \in \bigcap_{q=0}^{n-1} C_q^j \cap S$, if Carole chooses $e_j \in E$. Therefore it holds:

$$(x,y) \in R \triangleq \bigcup_{e_j \in E} \left(\bigcap_{q=0}^{n-1} C_q^j \cap S \right)$$

Now we will show that $\mu(R) \le 2^{-(n+1)}|E|$, if we choose A_q in the following way:

Definition 16.8. For each $q = 0, \ldots, n-1$ we define a collection of points x_i^q.

$$0 \le i \le 2^{q+1} : \; x_0^0 = 0,\; x_1^0 = 1 - \tfrac{1}{\sqrt{2}},\; x_2^0 = 1$$

$$q \ge 1 : \; x_i^q = \begin{cases} x_{\frac{i}{2}}^{q-1}, & \text{if } i \text{ is even} \\ x_{\frac{i+1}{2}}^{q-1} - \dfrac{x_{\frac{i+1}{2}}^{q-1} - x_{\frac{i-1}{2}}^{q-1}}{\sqrt{2}}, & \text{if } i \text{ is odd} \end{cases}$$

Remark 16.5. For all $q \in \{0, \ldots, n-1\}$ it holds: $x_0^q = 0$ and $x_{2^q+1}^q = 1$.

We set

$$A_q \triangleq \bigcup_{i=0}^{2^q-1} [x_{2i}^q, x_{2i+1}^q[\ \forall 0 \le q \le n-1$$

Theorem 16.1. *The preceding strategy is an optimal searching strategy that confines the pair (x, y) in a set, whose worst-case measure is $2^{-(n+1)}|E|$.*

In the parity testing model Carole thinks of two numbers (x, y) in $[0, 1[$. Paul chooses subsets $R \subset [0, 1[$ and asks questions like: 'Is either $x \in R \subset [0, 1[$ or $y \in R \subset [0, 1[$?' Again the search problem can be stated as a search in the search set $S = \{(x, y) : x, y \in [0, 1[, x > y\}$. We number Paul's questions from 0 to $n - 1$.

Definition 16.9.

1. We set $T_q \triangleq$ the subset Paul asks for in his q-th question.
2. We set

$$D_q = \begin{cases} T_q \times T_q \cup \bar{T}_q \times \bar{T}_q, & \text{if Carole answers 0 to the } q\text{-th question} \\ T_q \times \bar{T}_q \cup \bar{T}_q \times T_q, & \text{if Carole answers 1 to the } q\text{-th question} \end{cases}$$

Obviously $(x, y) \in D_q$, if Carole's answer was correct.

$$D_q^j \triangleq \begin{cases} D_q, & \text{if the } q\text{-th entry of } e_j \text{ is } 0 \\ \bar{D}_q, & \text{if the } q\text{-th entry of } e_j \text{ is } 1 \end{cases}$$

Obviously $(x, y) \in \bigcap_{q=0}^{n-1} D_q^j \cap S$, if Carole chooses $e_j \in E$. Therefore it holds:

$$(x, y) \in R \triangleq \bigcup_{e_j \in E} \left(\bigcap_{q=0}^{n-1} D_q^j \cap S \right)$$

Now we will show that $\mu(R) \le 2^{-(n+1)}|E|$, if we choose T_q as follows:

Definition 16.10. For each $q = 0, \ldots, n-1$ we define

$$T_q = \bigcup_{i=0}^{2^q-1} \left[\frac{2i}{2^{q+1}}, \frac{2i+1}{2^{q+1}} \right[$$

Theorem 16.2. *The preceding strategy is an optimal searching strategy that confines the pair (x, y) in a set, whose worst-case measure is $2^{-(n+1)}|E|$.*

17. Q-ary case

In this section we consider the problem of transmitting messages over a noisy q-ary channel with noiseless feedback. This problem is closely related to a sequential search with errors. Instead of using a binary coding alphabet, we introduce $\mathcal{Y} = \{0, \dots, q-1\}$ as the q-ary coding alphabet. All other variables and functions are chosen as before. The quiet-question-noisy-answer-game is changed in the following way. The sender and receiver have a common q-partition strategy. They divide the set of possible messages to q pairwise disjoint sets $(M_0, \dots, M_{q-1})$. The sender transmits the index of the set over the noisy channel. A "Devil" wants to avoid that the receiver gets the message by changing at most l answers. Again we consider the sets

$$S_j \triangleq \{x \in \mathcal{X} \, : \, \text{Paul get } (l-j) \text{ negative votes for } x\}.$$

Definition 17.1. The vector

$$\underline{v} = \big(|S_l|, |S_{l-1}|, \dots, |S_0|\big) = (v_l, v_{l-1}, \dots, v_0)$$

is referred to as a state (of the game). $\underline{v}$ is called a k-state if k questions are left.

Definition 17.2. Let $\underline{v} = \big(|S_l|, \dots, |S_0|\big)$ be an arbitrary state and let $M_0, \dots, M_{q-1}$ be the next partition used by Paul. We set $X_j^i = S_i \cap M_j$ and denote the question "Which subset contains the message?" by $[\underline{x}] = (x_{i,j})_{\substack{i=l\dots0\\ j=0\dots(q-1)}} = \begin{bmatrix} x_{l0} & \cdots & x_{l(q-1)} \\ \vdots & \ddots & \vdots \\ x_{00} & \cdots & x_{0(q-1)} \end{bmatrix}$, where $x_{i,j} \triangleq |X_j^i|$. $[\underline{x}]$ is called a question-matrix.

Definition 17.3. The state $\underline{s}$ can be reduced to the states $\underline{a}_1, \underline{a}_2, \dots, \underline{a}_q$ if there exists a question $[\underline{x}]$ such that $\underline{a}_1, \underline{a}_2, \dots, \underline{a}_q$ are possible states after this question.

Definition 17.4.

1. A 0-state $\underline{x}$ is called winning if $\sum_{i=0}^{l} x_i \leq 1$. Otherwise, it is called losing.
2. A k-state $\underline{x}$ is called winning if it can be reduced to q winning $(k-1)$-states. Otherwise, it is called losing.

3. A winning k-state is called borderline winning if it is a losing $(k-1)$-state.
4. A k-state is called singlet if $\sum_{i=0}^{k} |S_i| = 1$.
5. A k-state is called small if $\sum_{i=0}^{k} |S_i| \le q$.

Proposition 17.1.

1. *A winning n-state is also a winning k-state if $k > n$.*
2. *All borderline winning 1-states have the form $(0, \ldots, 0, a)$, where $a \le q$.*
3. *Let $\underline{x}$ be a winning k-state and $\underline{y}$ be some state. If $y_i \le x_i$ for all $i \le l$, then it is also a winning k-state.*

Definition 17.5. Let $\underline{x}$ be an arbitrary state. Then

$$V_n^q(\underline{x}) \triangleq \sum_{i=0}^{l} x_i \sum_{j=0}^{i} \binom{n}{j} (q-1)^j$$

is called the n-th volume of the state $\underline{x}$.

Proposition 17.2. *Let $\underline{x}$ be a state, which can be reduced to the states $\underline{a}_1, \underline{a}_2, \ldots \underline{a}_q$. Then*

$$V_n^q(\underline{x}) = \sum_{i=1}^{q} V_{n-1}^q(\underline{a}_i).$$

Proposition 17.3. *Let $\underline{x}$ be a winning n-state. Then*

$$V_n^q(\underline{x}) \le q^n.$$

Let $\mathcal{X} = \{1, \ldots, N\}$ be the set of messages and l be the maximal number of lies. We denote the minimal number of questions of Paul to find the searched number by using an optimal strategy by $L_l^q(N)$.

Corollary 17.4 (The Hamming bound). *Let $|\mathcal{X}| = N$ and $l \in \mathbb{N}$. Then*

$$L_l^q(N) = n \Longrightarrow N \le \frac{q^n}{\sum_{j=0}^{l} \binom{n}{j} (q-1)^j}.$$

Definition 17.6. Let $\underline{s}$ be an arbitrary state. Its character is defined as

$$ch(\underline{s}) \triangleq \min \left\{ k : V_k^q(\underline{s}) \le q^k \right\}.$$

Lemma 17.5. *Let $\underline{x}$ be a small state, but not a singleton state, $m \triangleq \max\{j : x_j \neq 0\}$ and $k \triangleq \begin{cases} \max\{j : x_j \neq 0 \text{ and } j \neq m\} & \text{if } x_m = 1, \\ m & \text{otherwise.} \end{cases}$*

If $z = k + m + 1$, then $\underline{x}$ is a borderline winning z-state.

Definition 17.7. Let $\underline{x} = \begin{pmatrix} x_k \\ \vdots \\ x_0 \end{pmatrix}$ be a state. Then $T(\underline{x}) = \underline{t} = \begin{pmatrix} t_{k-1} \\ \vdots \\ t_0 \end{pmatrix}$ with $t_i = x_{i+1}$ is called the translation of $\underline{x}$.

Theorem 17.1. *Let $\underline{x}$ be a state. If $\underline{x}$ is a winning n-state, then $T(\underline{x})$ is a winning $(n-2)$-state.*

Corollary 17.6. *Let $\underline{x}$ be a winning n-state. Then $T^m(\underline{x})$ is a winning $(n-2m)$-state.*

From this Corollary follows the following bound.

Theorem 17.2. *Let $N = q^i$ then $L_l^q(N) \geq \max\big\{i + 2l, ch\big((N, 0 \ldots, 0)\big)\big\}$.*

Aigner [12] conjectured that this bound is tight. It is shown in [29] that this conjecture is wrong, but the bound is not so bad.

Theorem 17.3. *Let $N, l, q \in \mathbb{N}$ with $q \geq 2$ then holds*

1. *If $\lceil \log_q N \rceil \leq q - 1$ then $L_l^q(N) = \lceil \log_q N \rceil + 2e$.*
2. *If $q - 1 < \lceil \log_q N \rceil \leq \min\{q(q-2), 2(q-1)\}$ then $\lceil \log_q N \rceil + 2e \leq L_l^q(N) \leq \lceil \log_q N \rceil + 2e + 1$.*

Aigner and Malinowski have solved independently the q-ary case if $l = 1$. They have the following equivalent results:

Theorem 17.4 (Aigner [12]). *Suppose that N and q are chosen as before and h, r are chosen in such a way that $N = hq - r$ and $0 \leq r < q$. Then*

$$L_1^q(N) = \begin{cases} \lceil \log_q N \rceil + 2 & \text{if } n \leq q^{q-1}, \\ \min\left\{k : \dfrac{q^k - r(k-1)(q-1)}{1 + k(q-1)} \geq N\right\} & \text{otherwise.} \end{cases}$$

Theorem 17.5 (Malinowski [57]). *Suppose that N and q chosen as before and h, r are chosen in such a way that $N = h(q+1) + r$ and $0 \le r < q+1$. Let $c = \min\left\{(q+1)^i \,:\, (q+1)^i \ge h\right\}$ and $r = 0$. Then*

$$L_1^q(N) = 1 + \begin{cases} ch(h, qh) & \text{if } h \ge (q+1)^{q-1}, \\ \max\left\{ch(h, qh), ch\big(c, c(h-1)\big)\right\} & \text{otherwise.} \end{cases}$$

Let $c = \min\left\{(q+1)^i \,:\, (q+1)^i \ge h+1\right\}$ and $r \ne 0$. Then

$$L_1^q(N) = 1 + \begin{cases} ch(h+1, qh+r-1) & \text{if } h \ge (q+1)^{q-1}, \\ \max\left\{ch(h+1, qh+r-1), ch\big(c, c(h-1)\big)\right\} & \\ & \text{otherwise.} \end{cases}$$

In general it is clear that the character is a lower bound for the minimal number of questions. In [29] this lower bound is improved for the case $l = 2$ and $N = q^i$ and it is shown that this bound can be attained.

Theorem 17.6. *Let $N = q^i$ with $i, q \in \mathbb{N}$ and not $q = i = 2$. Then*

$$L_2^q(N) = \min_{k \in \mathbb{N}} \left\{k \,:\, V_k^q(q^i, 0, 0) \le q^k \text{ and } V_{k-2}^q(0, q^i, 0) \le q^{k-2}\right\}.$$

If we do not restrict N we have the following result.

Theorem 17.7. *Let $c = \min_{k \in \mathbb{N}}\left\{k \,:\, V_k^q(q^i, 0, 0) \le q^k \text{ and } V_{k-2}^q(0, q^i, 0) \le q^{k-2}\right\}$ and $\lfloor \log_q N \rfloor \ge q-1$, then $c \le L_2^q(N) \le c+1$.*

18. The probabilistic case

Instead of giving a special number of lies one could consider the following probabilistic problem. Carole lies with the probability p. Without feedback this model is equivalent to the well known binary symmetric channel (BSC) for a $(\mathcal{X}, p)$ channel. In this model we have a binary input alphabet $\mathcal{X}$ and a binary output alphabet $\mathcal{Y}$. The input symbols are complemented with probability p. For this channel we make the following definitions:

Definition 18.1. Let $\mathcal{X} = \{0, 1\}$ be the binary coding alphabet.

1. An (M, n, p) code for a BSC consists of the following:
   (a) A message set $\mathcal{M} = \{1, \dots, M\}$.
   (b) An encoding function $c^n : \mathcal{M} \to \mathcal{X}^n$.
   (c) A decoding function $d : \mathcal{X}^n \to \mathcal{M}$.
1. The error probability of a (M, N, p) code is defined as:

$$\lambda^n = \max_{i \in \mathcal{M}} P\big(g(Y^n) \neq i \mid X^n = c^n(i)\big).$$

2. The rate R of an (M, n, p) code is $R = \frac{\log M}{n}$.
3. A rate R is achievable if there exists a sequence of (M, n, p) codes such that the error probability tends to 0 as $n \to \infty$.
4. The capacity C for a BSC is the supremum of all achievable rates.

For a discussion of shortcomings of this definition for general channels see [3].

It is well known that the capacity of the BSC is $1 - h(p)$. The generalization is the discrete memoryless channel (DMC). In this model we consider arbitrary input and output alphabets and a transmission matrix with the transmission probabilities. The known proofs of the coding theorem for DMC use either random coding method or a maximal coding method. Shannon proved that feedback can not increase the capacity of a DMC, but it provides new possibilities for the construction of codes. The first attempt was made by Horstein. In 1971 Ahlswede [1] gave a constructive proof of the coding theorem for discrete memoryless channels with feedback. The result can be found in section 19. In the paper [67] the author considered the case if Carole lies with probability p independently in every step of the game. He considered three models:

1. The continuous case: Carole thinks of a real number in the interval $[0, 1]$. The goal of Paul is to find an interval of length ε which contains this number.
2. The discrete case: Original Game where Carole thinks of N numbers.
3. The discrete unbounded problem: Carole thinks of a natural number.

The author found out for which values of p the game is feasible. He got the following result.

Theorem 18.1. *Let Carole lie with probability p independently in every step of the game, $p \in (0, 1)$, $\varepsilon > 0$, $r \leq \frac{1}{\varepsilon}$ and $N \in \mathbb{N}$.*

1. *Paul wins version 1 of the game if and only if* $p < \frac{1}{2}$. *And there exists a winning strategy in* $O(\log r)$ *steps.*
2. *Paul wins version 2 of the game, if* $p < \frac{1}{3}$. *And there exists a winning strategy in* $O(\log N)$ *steps.*
3. *Paul wins version 3 of the game, if* $p < \frac{1}{3}$.

19. Search and information theory

In the previous sections we have seen that searching with lies is equivalent to coding with feedback. In this section we will consider the discrete memoryless channel (DMC), which is a generalization of the BSC. With $\mathcal{X}$ and $\mathcal{Y}$ we denote the input and output alphabets. The channel is fully characterized by specifying the transmission matrix w, where $1 \geq w(y|x) \geq 0$ for all $x \in \mathcal{X}$, $y \in \mathcal{Y}$ and $\sum_{y \in \mathcal{Y}} w(y|x) = 1$ for all $x \in \mathcal{X}$. We assume the independence structure $P(Y^n = y^n|x^n) = \prod_{1 \leq t \leq n} w(y_t|x_t)$ for $x^n \in \mathcal{X}^n$ and $y^n \in \mathcal{Y}^n$. $P(Y^n = y^n|x^n)$ is the probability that the randomly received sequence Y^n is equal to y^n if x^n was transmitted. We also assume noiseless feedback and call such a channel DMCF. In [1] one can find a constructive proof of the coding theorem for the DMCF. The entropy $H(X) = H(P) = -\sum_{x \in X} P(X = x) \log P(X = x)$ is often interpreted as a measure of the uncertainty about which value X assumes. Let $Q = P \cdot W$ denote the distribution of the output variable Y, then the mutual information $I(P, W) = I(X \wedge Y) = H(X) - H(X|Y)$ measures the information which we obtain about X if Y is observed. In [1] with the help of list codes a concrete meaning is assigned to this interpretation. An (N, l, λ, L)-list code is a set of pairs (u_i, D_i) with $u_i \in \mathcal{X}^l$, $D_i \subset \mathcal{Y}^l$, $W(D_i|u_i) \geq 1 - \lambda$ for all $1 \leq i \leq N$ and $\sum_{1 \leq i \leq N} 1_{D_i}(y^l) \leq L$ for all $y^l \in \mathcal{Y}^l$.

Lemma 19.1 (Ahlswede's list reduction lemma). *There exists an* (N, l, λ, L)*-list code with the following properties:*

1. $N \approx \exp\left\{H(X)l\right\}$,
2. $L \approx \exp\left\{H(X|Y)l\right\}$.

The exact notions in this lemma depend on the kind of typical sequences used for the construction of the list code. By iteratively applying this lemma we get rates which approach the capacity C arbitrary closely. This method of list reduction is generalized in [2] for the arbitrarily varying channel

(AVC). Let $\mathcal{W} = \{w(\cdot|\cdot|s) : s \in S\}$ be a finite set of stochastic $|\mathcal{X}| \times |\mathcal{Y}|$ matrices. For each $s^n \in S^n$ we define $W(y^n|x^n|s^n) = \prod_{1 \le t \le n} w(y_t|x_t|s_t)$ and the sets $\mathcal{W}^n := \{W(\cdot|\cdot|s^n) : s \in S\}$. An AVC is then given by the sequence $(\mathcal{W}^n)_{n=1}^{\infty}$. We denote by $\overline{\mathcal{W}}$ the convex hull of $\mathcal{W}$ and by $\overline{\overline{\mathcal{W}}}$ its row-convex hull.

Lemma 19.2 (Ahlswede's list reduction lemma for AVC). *There exists an (N, l, λ, L)-list code for the AVC $\mathcal{W} = \overline{\overline{\mathcal{W}}}$ with the following properties:*

1. $N \approx \exp\{H(X)l\}$,
2. $L \approx \exp\{l \max\{H(X|Y(w) : w \in \mathcal{W})\}$.

Like before by iteratively applying this lemma it is shown that the capacity has the formula $C_F(\overline{\overline{\mathcal{W}}}) = \max_{P \in \mathcal{P}(\mathcal{X})} \min_{\mathcal{W} \in \overline{\overline{\mathcal{W}}}} I(P, W)$, if the capacity is positive. In [2] it is shown that the coding problem for the DMC with feedback is in the error-free case a special case of the coding problem for the AVC with feedback. We obtain a formula, which was conjectured by Shannon in [73].

Later in the paper [10] the capacity for the AVC $\mathcal{W} = \overline{\overline{\mathcal{W}}}$ with feedback is determined completely, that is without the positivity assumption. More importantly the authors also solve the problem for general $\mathcal{W}$. The formula distinguishes three cases and therefore the authors speak of a trichotomy. To establish this theorem the authors use the list reduction lemma, the balanced coloring lemma ([4]) and techniques of identification theory ([5]).

Theorem 19.1 (trichotomy theorem). *Let $\hat{\mathcal{W}} = \{\hat{W} : \hat{W}(\cdot|x) \in \hat{\mathcal{W}}(x)$ for all $x \in \mathcal{X}$ and $\hat{W}(y|x) \in \{0,1\}$ for all $y \in \mathcal{Y}\}$, where $\hat{\mathcal{W}}(x) = \{W(\cdot|x,s) : s \in S\}$ and let $C_R(\mathcal{W}) = \max_P \min_{W \in \overline{\mathcal{W}}} I(P, W)$ denote the random code capacity.*

$$C_F(\mathcal{W}) = \begin{cases} 0, & \text{iff } C_R(\mathcal{W}) = 0 \text{ or } \mathcal{Y}_x \cap \mathcal{Y}_{x'} \neq \emptyset \ \forall x, x' \in \mathcal{X} \\ C_R(\mathcal{W}), & \text{iff } C_R(\mathcal{W}) > 0 \text{ and } \mathcal{Y}_x = \emptyset \text{ for some } x \\ \min\{C_R(\mathcal{W}), C_F(\hat{\mathcal{W}})\}, & \\ & \text{iff } C_R(\mathcal{W}) > 0 \text{ and } \mathcal{Y}_x \neq \emptyset \text{ for all } x \end{cases}$$

The AVC models a robust search problem in the presence of noise. There exist many papers about multi-way-channels with feedback, mostly with partial results. For instance for the two-way-channel in the pioneering paper of Shannon [74] the capacity region is still unknown.

20. Identification and a general search model

In the first part of the paper [8] Ahlswede describes a general communication model which includes identification as a special case. We will formulate this model as a search model.

One can conceive of many situations in which Paul has different goals.

A nice class of such situations can, abstractly, be described by a family $\Pi(\mathcal{M})$ of partitions of $\mathcal{M}$. Paul $\pi \in \Pi(\mathcal{M})$ wants to know only which member of the partition $\pi = (A_1, \dots, A_r)$ contains m, the true number, which is known to Carole.

We describe now some seemingly natural families of partitions.

Model 1: $\Pi_S = \{\pi_{sh}\}$, $\pi_{sh} = \big\{ \{m\} \,:\, m \in \mathcal{M} \big\}$. This describes the classical problem.

Model 2: $\Pi_I = \{\pi_m \,:\, m \in \mathcal{M}\}$ with $\pi_m = \big\{ \{m\}, \mathcal{M} \backslash \{m\} \big\}$. Paul π_m wants to know whether m occurred or not. This is the identification problem introduced in [6].

Model 3: $\Pi_K = \big\{ \pi_S \,:\, |S| = K,\ S \subset \mathcal{M} \big\}$ with $\pi_S = \{S, \mathcal{M} \backslash S\}$. This is an interesting generalization of the identification problem. It is called K–identification.

Model 4: $\Pi_R = \Big\{ \pi_r \,:\, \pi_r = \big\{ \{1, \dots, r\}, \{r+1, \dots, M\} \big\},\ 1 \leq r \leq M-1 \Big\}$. Here Paul π_r wants to know whether the number exceeds r or not. We speak of the ranking problem.

Model 5: $\Pi_B = \{\Pi_A \,:\, A \subset \mathcal{M}\}$. Here $\pi_A = \{A, \mathcal{M} \backslash A\}$ wants to know the answer to the binary question "Is m in A?".

Model 6: $\mathcal{M} = \{0,1\}^\ell$, $\Pi_C = \{\pi_t \,:\, 1 \leq t \leq \ell\}$ with $\pi_t = \Big\{ \big\{ (x_1, \dots, x_\ell) \in \mathcal{M} \,:\, x_t = 1 \big\}, \big\{ (x_1, \dots, x_\ell) \in \mathcal{M} \,:\, x_t = 0 \big\} \Big\}$. Decoder π_t wants to know the t–th component of the vector valued message $(x_1, \dots, x_\ell)$.

21. The game without feedback

If we consider the Ulam–Rényi game without feedback, all strategies are equivalent to error correcting codes. Thus we need some definitions for error correcting codes.

Definition 21.1.

1. We call a code a (N, l, n) code, if we can transmit N messages with a block code of length n if at most l errors happen.
2. We denote by $N(l, n)$ the maximal number N of messages such that a (N, l, n) code exists.

Our first goal is to compare the number of questions of the adaptive game and the non-adaptive game.

Definition 21.2. We denote by $G_l(N)$ the minimal number of questions of Paul to find the searched number by using an optimal non-adaptive strategy.

In the paper [43] the authors make such comparisons. It is clear that $L_l(N) \leq G_l(N)$. Thus the Hamming Bound is also a lower bound for the non-adaptive game. The Hamming bound is known much longer for the non-adaptive case. In [62] it is shown that $G_1(2^{20}) = 25$. In [79] the following is shown.

Theorem 21.1.

1. $G_1(2^{20}) = 25$
2. $G_2(2^{20}) \in \{29, 30\}$
3. $G_3(2^{20}) \in \{33, 34, 35\}$
4. $G_4(2^{20}) \in \{37, 38, 39, 40\}$

In [13] Aigner considered the q-ary case without feedback. We will use the notation $G_l^q(N)$ for the minimal number of questions of Paul to find the searched number by using an optimal non-adaptive q-ary strategy. For one lie he gets the following result.

Theorem 21.2. *Let q be a prime number then holds*

1. $L_1^q(N) = G_1^q(N)$ *if* $n \leq q^{q-1}$.
2. $L_1^q(N) \leq G_1^q(N) \leq L_1^q(N) + 1$ *if* $n > q^{q-1}$.

Further results about error-correcting codes can be found in [56].

22. Learning and searching

Mundici and Cicalese pointed out in [24] that learning and searching are strongly connected. In the Ulam–Rényi game at every step Paul learns something about Carole's secret element.

Let us consider a learning model which is strongly connected to the Ulam–Rényi game. We have a teacher and a student. There is a collection of concepts $R_1, \ldots, R_m$ and a set of words $\mathcal{X} = \{1, \ldots, M\}$. Each concept R_j is a subset of the words. The concepts are not disjoint, but all concepts are different sets. The teacher generates a sequence of length n of words referring to a concept R_i. The student knows that the teacher speaks about one of m concepts and he knows which words are used to describe concepts. He learns the concept in the following way. In step k he reads the k-th word of the sequence and makes a binary test to learn the concept of the teacher. He is able to test if $x \in \bigcup_{i \in I} R_i$, where $I \subset \{1, \ldots, m\}$. The test may be faulty. This means that at most l tests are wrong. If we choose $\mathcal{X} = \{1, \ldots, M\}$, $R_j = \{\pi(j)\}$ for $j = 1, \ldots, M = m$, where π is a permutation of $\mathcal{X}$ and set for the teacher-sequence $(t_j)_{j=1}^n$, $t_j = x \in R_i$ for all j. Obviously this is equivalent to the Ulam–Rényi game. There exist several variants of this model. We can restrict the tests, the teacher-sequence may be also faulty or the tests are wrong with a given probability.

In [50] the authors consider a probabilistic variant of the learning model. In their setting no concept is a subset of another and the tests are not faulty. In first step the learner makes a guess of the concept of the teacher by choosing one concept at random and checks for consistency with the first word of the teacher-sequence. He is sticking to his choice at each step until it is proven wrong. They consider two kinds of learning strategies: The memoryless learner, who chooses a new concept at random from all concepts if one test is wrong and the learner with full memory, who never chooses a concept again, if it has been rejected before. All random choices are done according to the uniform distribution. They get the following result.

Theorem 22.1. *Let $N(p)$ be the number of steps it takes for the student to have probability $1-p$ of learning the concept. There exists constants c_1, c_2, c_3, c_4 such that*

$$\lim_{n \to \infty} P\left(c_1 < \frac{N(p)}{|\log(p)|\, n \log(n)} < c_2\right) = 1$$

holds for the memoryless learner and

$$\lim_{n\to\infty} P\left(c_3 < \frac{N(p)}{\left|(1-p)^2\right| n\log(n)} < c_4\right) = 1$$

holds for the learner with full memory.

Many models of this kind for learning can be found in [52], [51] and [49].

23. Unequal protection with feedback

In this section we will consider a modified channel model. A detailed version of this section can be found in [7]. We consider a communication system, with one sender (or encoder) E and two receivers (or decoders), D_1 and D_2. The sender E wants to send a message $i \in \mathcal{M}_1$ to D_1 and a message $j \in \mathcal{M}_2$ to D_2 simultaneously and he encodes each pair (i, j) of messages into a binary sequence of length n. This sequence is sent to the two receivers via two independent channels. Because of the noise in the channels, the output sequences received by the two receivers may have errors at most at $t_1 = n\tau_1$ and $t_2 = n\tau_2$ bits respectively. Moreover we assume that we have noiseless feedback. That means, the sender is allowed to choose the t-th bit of input according to the first $(t-1)$ bits of both output sequences. We call the code a binary $\big(n, M_1, M_2, (t_1, t_2)\big)$ feedback code (or briefly a fb-code) for broadcast channel if $|\mathcal{M}_k| = M_k$ $k = 1, 2$. A codeword is in the form of: $c(i, j) = \left(c_1(i, j), \ldots, c_n(i, j, y_1^{n-1}, y_2^{n-1})\right)$, where c_i : $\mathcal{M}_1 \times \mathcal{M}_2 \times Y_1^{i-1} \times Y_2^{i-1} \to Y$ is a function for the i-th code letter which depends on the message we want to transmit and the $(i-1)$ bits which have been received by Receiver 1 and 2. We also call c a common strategy (of the Sender and the Receivers). We will consider the special case, if $t_1 = 0$. To present the bound we need an isoperimetric inequality. For a subset $A \subset \{0,1\}^n$, let $\Gamma^t(A) = \big\{x^n$: there exists an $a^n \in A$ such that $d_H(x^n, a^n) \le t\big\}$ for $1 \le t \le n$. Then the isoperimetric problem for binary Hamming space asks what sets achieve $\min_{A:\,|A|=u} \left|\Gamma^t(A)\right|$ for all k. This problem was solved by Harper [41] and Katona [46]. They showed that the initial segments of following order are always optimal for the minimization. Let x^n and y^n be two binary sequences with the same Hamming weight. Then we say that x^n precedes y^n in the squashed order if $y_i = 1$ at the largest component i where $x_i \ne y_i$. A binary sequence x^n precedes a binary sequence y^n in Tapper order, briefly H-order, if the Hamming weight of x^n is less than the the Hamming weight of y^n or they have the same Hamming weight and $1^n - x^n$ precedes $1^n - y^n$ in the squashed order. In the binary Hamming space of dimension n the uth initial segment in H-order is denoted by $S_{n,u}$. One can verify that

$$\Gamma^t(S_{n,u}) \subset \Gamma^t(S_{n,v}) \tag{23.1}$$

for all n, t if $u < v$. We present the outer bound in terms of function G which was introduced by Katona [46]. For given n any non-negative integer u can be uniquely represented as

$$u = \binom{n}{n} + \cdots + \binom{n}{k+1} + \binom{\alpha_k}{k} + \ldots + \binom{\alpha_t}{t} \tag{23.2}$$

with $n > \alpha_k > \cdots > \alpha_t \geq t \geq 1$. Then the function G is defined as

$$G(n,u) = \binom{n}{n} + \binom{n}{n-1} + \cdots + \binom{n}{k} + \binom{\alpha_k}{k-1} + \ldots \binom{\alpha_t}{t-1}. \tag{23.3}$$

Moreover we rewrite G as $G^{\circ 1}$ and define $G^{\circ t}(n,\cdot) = G\big(n, G^{\circ t-1}(n,\cdot)\big)$ recursively. The Isoperimetric Theorem says that

$$\min_{A\,:\,|A|=u} \big|\Gamma^t(A)\big| = \big|\Gamma^t(S_{n,u})\big| = G^{\circ t}(n,u). \tag{23.4}$$

To show (23.4) Katona proved that if $0 \leq u_1 \leq u_0$ and $u = u_0 + u_1$,

$$G(n,u) \leq \max\big[u_0,\ G(n-1,u_1)\big] + G(n-1,u_0). \tag{23.5}$$

To obtain the outer bound we need its generalization.

Lemma 23.1. *Let u_0, u_1 be non-negative integers and $u = u_0 + u_1$. Then*

$$\begin{aligned} G^{\circ t}(n,u) \leq\ & \max\big[G^{\circ t}(n-1,u_0),\ G^{\circ t-1}(n-1,u_1)\big] \\ & + \max\big[G^{\circ t-1}(n-1,u_0),\ G^{\circ t}(n-1,u_1)\big] \end{aligned} \tag{23.6}$$

for all n and $0 \leq t \leq n$.

Let us now turn to our problem of binary error correcting code with feedback for broadcast channels. We shall obtain the outer bound by counting the number of possible output sequences. Consider a family of encoding functions of a binary feedback code $f_m^n : \{0,1\}^{n-1} \longrightarrow \{0,1\}^n$, $m \subset \mathcal{M}$,

$$(23.7)\quad f_m^n(y^{n-1}) = \big(f_m^{(1)}, f_m^{(2)}(y_1), f_m^{(3)}(y_1,y_2), \dots, f_m^{(n)}(y_1,y_2,\dots,y_{n-1})\big).$$

When they are input into a channel (of one receiver) with noise of binary additive errors and noiseless feedback, the output $y^n = (y_1, \dots, y_n)$ with

$$(23.8)\quad y_1 = f_m^{(1)} + e_1 \text{ and } y_t = f_m^{(t)}(y_1, y_2, \dots y_{t-1}) + e_t \text{ for t} = 2,3, \dots, \text{n}.$$

is uniquely determined by the encoding function f_m^n and the binary additive noise $e^n = (e_1, e_2, \dots, e_n)$ occurring in the transmission and so can be regarded as a function $\Phi(f_m^n, e^n)$. For a family of encoding functions $\{f_m^n : m \in \mathcal{M}\}$, and a set $\mathcal{E}$ of error patterns, we write

$$(23.9)\quad \Phi(f_{\mathcal{M}}, \mathcal{E}) = \big\{y^n : \text{ there exist } m \in \mathcal{M} \text{ and } e^n \in \mathcal{E} \text{ such that } y^n = \Phi(f_m^n, e^n)\big\}.$$

We believe that the following lemma is independently interesting in Combinatorics because it can be considered as an isoperimetric theorem for the sequences generated with feedback. By choosing $\mathcal{E} = \mathcal{E}(n,t)$ the set of binary sequences of length n with Hamming weight not exceeding t, as the set of error patterns then we have

Lemma 23.2. *For any family $\{f_m^n : m \in \mathcal{M}\}$ of encoding functions,*

$$(23.10)\quad \big|\Phi\big(f_{\mathcal{M}}^n, \mathcal{E}(n,t)\big)\big| \geq G^{ot}\big(|\mathcal{M}|\big).$$

With this lemma we can obtain the following theorem.

Theorem 23.1 (Hamming Bound for fb-Codes Correcting $(0,t)$-Error for Broadcast Channels)**.** *If there exists an $\big(n, M_1, M_2, (0,t)\big)$ binary fb-codes for broadcast channels,*

$$(23.11)\quad M_2 \leq \frac{2^n}{G^{ot}(n, M_1)}$$

If we consider the case $\frac{t_2}{n} \to \tau$ if $n \to \infty$, we obtain the following.

Corollary 23.3. *Let p be so chosen that $R_1 = h(p)$. Then $R_2 \leq 1 - h(p+\tau)$.*

It is very easy to verify that one can obtain better rates in this model than in the model without feedback (by using timesharing).

Acknowledgments.

The author is grateful to Rudolf Ahlswede, Lars Bäumer, Ferdinando Cicalese, Daniele Mundici, Andrzej Pelc and the two unknown referees for useful comments and helpful discussions.

References

[1] R. Ahlswede, A constructive proof of the coding theorem for discrete memoryless channels with feedback, in: *Transactions of the Sixth Prague Conference on Information Theory, Statistical Decision Functions, Random Processes, Tech. Univ., Prague* (1973), pp. 39–50.

[2] R. Ahlswede, Channels with arbitrarily varying channel probability functions in the presence of noiseless feedback, *Z. Wahrsch. th. u. verw. Geb.,* **25** (1973), 239–252.

[3] R. Ahlswede, Concepts of performance parameters for channels, General Theory of Information Transfer and Combinatorics, *Lecture Notes in Computer Science,* Vol. **4123** (2006).

[4] R. Ahlswede, Elimination of correlation in random codes for AVC, *Z. Wahrsch. th. u. verw. Geb.,* **44** (1978), 159–175.

[5] R. Ahlswede and G. Dueck, Identification in the presence of feedback – a discovery of new capacity formulas, *IEEE Trans. Inform. Theory,* **35**, no. 1 (1989), 30–36.

[6] R. Ahlswede and G. Dueck, Identification via channels, *IEEE Trans. Inform. Theory,* Vol. **35** (1989), 15–29.

[7] R. Ahlswede, N. Cai and C. Deppe, An isoperimetric theorem for sequences generated by feedback and feedback codes for unequal protection of messages, *Problems in Information Transmission,* Vol. **37**, No. 4 (2001), 332–338.

[8] R. Ahlswede, General Theory of information transfer, *Preprint* 97-118, SFB 343, Universität Bielefeld, 1997.

[9] R. Ahlswede and I. Wegener, Suchprobleme, *Teubner* (1979), English translation: *Wiley* (1987), Russian translation: *MIR* (1982).

[10] R. Ahlswede and N. Cai, The AVC with noiseless feedback and maximal error probability: a capacity formula with a trichotomy, in: *Numbers, Information and Complexity (Bielefeld, 1998),* Kluwer Acad. Publ. (Boston, MA, 2000), pp. 151–176.

[11] M. Aigner, Combinatorial Search, *Wiley-Teubner* (1988).

[12] M. Aigner, Ulams Millionenspiel, *Math. Semesterber.,* vol. **42** (1995), 71–80.

[13] M. Aigner, Searching with lies, *J. Combin. Theory Ser. A,* vol. **74** (1996), 43–56.

[14] J. A. Aslam and A. Dhagat, Searching in the presence of linearly bounded errors, *Proceedings of the 23rd Annual ACM Symposium of Theory of Computing* (1991).

[15] V. B. Balakirsky, A direct approach to searching with lies, *Preprint 98-070, University of Bielefeld* (1998).

[16] R. Beigel, Unbounded searching algorithms, *SIAM J. Comput.*, Vol. **19**, no. 3 (1990), 522–537.

[17] J. L. Bentley and C. C.-C. Yao, An almost optimal algorithm for unbounded searching, *Inform. Process. Lett.*, **5** (1976), 82–87.

[18] E. R. Berlekamp, Block coding for the binary symmetric channel with noiseless, delayless feedback, in: *H. B. Mann, Error Correcting Codes, Wiley* (1968), pp. 61–85.

[19] E. R. Berlekamp, R. Hill and J. Karim, The solution of a problem of Ulam on searching with lies, *Proc. IEEE Int. Symp. on Inform. Theory* (MIT, Cambridge, MA USA, 1998), p. 244.

[20] A. de Bonis, L. Gargano and U. Vaccaro, Group testing with unreliable tests, *Inf. Sci.,* vol. **96**, no. 1–2 (1997), 1–12.

[21] M. V. Burnashev, Data transmission over a discrete channel with feedback, *Problems Inform. Transmission,* vol. **12** (1976), 250–265, translated from *Problemy Peredachi Informatsii,* vol. **12**, no. 4 (1976), 10–30 (in Russian).

[22] F. Cicalese, Reliable computation with unreliable information, *PhD thesis,* Univ. of Salerno, Dept. of Computer Science (2000).

[23] F. Cicalese and U. Vaccaro, An improved heuristic for the Ulam–Rényi game, *Inf. Proc. Letters,* Vol. **73** (2000), 119–124.

[24] F. Cicalese and D. Mundici, Learning and the Art of Fault–Tolerant Guesswork, *Handbook Chapter, In: Perspectives on Adaptivity and Learning. Stamatescu, I. et al., Eds., Springer-Verlag* (2003), pp. 115–140.

[25] F. Cicalese, D. Mundici and U. Vaccaro, Least adaptive optimal search with unreliable tests, *Theor. comput. Sci.,* vol. 270, no. 1–2 (2002), 877–893.

[26] F. Cicalese and D. Mundici, Perfect two-fault tolerant search with minimum adaptiveness, *Adv. Appl. Math.* **25**, no. 1 (2000), 65–101.

[27] F. Cicalese and D. Mundici, Optimal binary search with two unreliable tests and minimum adaptiveness, *Lect. Notes Comput. Sci.,* **1643** (1999), 257–266.

[28] F. Cicalese and D. Mundici, Optimal coding with one asymmetric error: below the sphere packing bound, *Lecture Notes in Comput. Sci.,* **1858** (2000), 159–169.

[29] F. Cicalese and U. Vaccaro, Optimal strategies against a liar, *Theoretical Computer Science,* Vol. **230** (2000), 167–193.

[30] F. Cicalese, D. Mundici and U. Vaccaro, Rota-Metropolis cubic logic and Ulam–Rényi games, in: *Algebraic combinatorics and computer science: a tribute to Gian-Carlo Rota,* Springer-Italia (Milan, 2001), pp. 197–244.

[31] F. Cicalese and D. Mundici, Optimal Searching Strategies for two Asymmetric Lies, *preprint* (2002).

[32] J. Czyzowicz, D. Mundici and A. Pelc, Solution of Ulam's problem on binary search with two lies, *J. Combin. Theory Ser. A,* vol. **49** (1988), 384–388.

[33] J. Czyzowicz, D. Mundici and A. Pelc, Ulam's searching game with lies, *J. Combin. Theory Ser. A*, vol. **52** (1989), 62–76.

[34] J. Czyzowicz, K. B. Lakshmanan and A. Pelc, Searching with a forbidden lie pattern in responses, *Inform. Process. Lett.*, vol. **37** (1991), 127–132.

[35] C. Deppe, Strategies for the Rényi–Ulam Game with fixed number of lies, *Theoretical Computer Science,* **314** (2004), 45–55.

[36] C. Deppe, Solution of Ulam's searching game with three lies or an optimal adaptive strategy for binary three-error-correcting codes, *Discrete Math.,* **224**, no. 1–3 (2000), 79–98.

[37] D. DesJardins, Precise coding with noiseless Feedback, *PhD thesis,* University of California at Berkeley, Department of Mathematics (2001).

[38] A. Dhagat, P. Gacs and P. Winkler, On playing twenty questions with a liar, *BUCS Tech Report #91-006, Boston University* (1992).

[39] D.-Z. Du and F. K. Hwang, Combinatorial Group Testing, World Scientific (Singapore, 1993).

[40] W. Guzicki, Ulam's searching game with two lies, *J. Combin. Theory Ser. A,* vol. **54** (1990), 1–19.

[41] L. H. Harper, Optimal numberings and isoperimetric problems on graphs, *J. Combinatorial Theory,* **1** (1966), 385–393.

[42] R. Hill, Searching with lies, *Surveys in Combinatorics, Lecture Note Series,* **218** (1995), 41–70.

[43] R. Hill and J. P. Karim, Searching with lies: the Ulam problem, *Discrete Math.*, vol. **106/107** (1992), 273–283.

[44] M. Horstein, Sequential transmission using noiseless feedback, *IEEE Trans. Inform. Theory,* vol. **9**, no. 3 (1963), 136–142.

[45] J. P. Karim, Searching with lies: the Ulam problem, *PhD thesis,* University of Salford, Department of Mathematics and Computer Science.

[46] G. Katona, A theorem of finite sets, in: *Theory of graphs* (Proc. Colloq., Tihany, 1966), Academic Press (New York, 1968), pp. 187–207.

[47] D. J. Kleitman, A. R. Meyer, R. L. Rivest, J. Spencer and K. Winklmann, Coping with errors in binary search procedures, *J. Comput. System Sci.,* vol. **20** (1980), 396–404.

[48] D. E. Knuth, Supernatural numbers, *The Mathematical Gardner,* ed. D. A. Klarner, Wadsworth International (1981), pp. 310–325.

[49] N. L. Komarova, P. Niyogi and M. A. Nowak, Evolution of universal grammar, *Scienc,e* **291** (2001), 114–118.

[50] N. L. Komarova and I. Riven, Mathematics of Learning, *Arxiv, Preprintserver,* math.PR/0105235, (http://arxiv.org), 2001.

[51] N. L. Komarova and M. A. Nowak, Natural selection of the critical period for grammar acquisition, *Proc. Royal Soc. B.,* **268**, No. 1472 (2001), 1189–1196.

[52] N. L. Komarova, P. Niyogi and M. A. Nowak, The evolutionary dynamics of grammar acquisition, *J. Theor. Biology* **209** (2001), 43–59.

[53] K. B. Lakshmanan and B. Ravikumar, Coping with known patterns of lies in a search game, *Theoret. Comput. Sci.,* vol. **33** (1984), 85–94.

[54] E. L. Lawler and S. Sarkissian, An algorithm for Ulam's game and its application to error correcting codes, *Inform. Process. Lett.,* vol. **56** (1995), 89–93.

[55] K. Leweling, Codierung eines diskreten gedächtnislosen Kanals mit Rückkopplung, *Diploma Thesis, Fakultät für Mathematik, Universität Bielefeld* (1975).

[56] F. J. MacWilliams and N. J. A. Sloane, The theory of error-correcting codes, II. North-Holland Mathematical Library, Vol. **16**. North-Holland Publishing Co. (Amsterdam–New York–Oxford, 1977).

[57] A. Malinowski, K-ary searching with a lie, *ARS Combin.,* vol. **37** (1994), 301–308.

[58] D. Mundici, The complexity of adaptive error-correcting codes, *Springer Lecture Notes in Computer Science,* **533** (1991), 300–307.

[59] D. Mundici and A. Trombetta, Optimal comparison strategies in Ulam's searching game with two lies, *Theoretical Computer Science,* vol. **182** (1997), 217–232.

[60] A. Negro and M. Sereno, Ulam's searching game with three lies, *Adv. in Appl. Math.,* **13** (1992), 404–428.

[61] A. Negro and M. Sereno, Solution of Ulam's problem on binary search with three lies, *J. Combin. Theory Ser. A,* vol. **59** (1992), 149–154.

[62] I. Niven, Coding theory applied to a problem of Ulam, *Math. Mag.,* vol. **61** (1988), 275–281.

[63] A. Pelc, Coding with bounded error fraction, *Ars Combinatoria,* vol. **24** (1987), 17–22.

[64] A. Pelc, Detecting errors in searching games, *J. Combin. Theory Ser. A,* vol. **51** (1989), 43–54.

[65] A. Pelc, Detecting a counterfeit coin with unreliable weighings, *Ars Combinatoria,* vol. **27** (1989), 181–192.

[66] A. Pelc, Solution of Ulam's problem on searching with a lie, *J. Combin. Theory Ser. A,* vol. **44** (1987), 129–140.

[67] A. Pelc, Searching with known error probability, *Theoret. Comput. Sci.,* **63** (1989), 185–202.

[68] A. Pelc, Searching games with errors – fifty years of coping with liars, *Theoret. Comput. Sci.,* **270** (2002), 71–109.

[69] A. Rényi, A Diary on Information Theory, *Wiley* (1987), translation of Napló az információelméletről (1976).

[70] A. Rényi, On a problem of information theory, *MTA Mat. Kut. Int. Közl.,* **6B** (1961), 505–516.

[71] G. Rong and L. Yizhong, Generalized Solution of Ulam's Problem, *Chinese Journal of Contemporary Mathematics,* vol. **13** (1992), 323–331.

[72] J. P. M. Schalkwijk, A class of simple and optimal strategies for block coding on the binary symmetric channel with noiseless feedback, *IEEE Trans. Inform. Theory,* vol. **17** (1971), no. 3.

[73] C. E. Shannon, The zero-error capacity of a noisy channel, *IRE Trans. Inform. Th.,* **3** (1956), 3–15.

[74] C. E. Shannon, Two-way communication channels, *Proc. 4th Berkeley Sympos. Math. Statist. and Prob.*, Vol. I (1961), pp. 611–644.

[75] J. Spencer, Guess a number – with lying, *Math. Mag.,* vol. **57**, no. 2 (1984), 105–108.

[76] J. Spencer, Ulam's searching game with a fixed number of lies, *Theoret. Comput. Sci.,* vol. **95** (1992), 307–321.

[77] J. Spencer and P. Winkler, Three thresholds for a liar, *Combin. Probab. Comput.,* vol. **1** (1992), 81–93.

[78] S. Ulam, *Adventures of a Mathematician,* Scribner (NY, 1976).

[79] T. Verhoeff, An update table of minimum distance bounds for binary codes, *IEEE Trans. Inform. Theory,* vol. **33** (1987), 665 680.

[80] T. Veugen, Multiple-repetition coding for channels with feedback, *PhD thesis,* Eindhoven University of Technology (1992).

[81] K. Sh. Zigangirov, Number of correctable errors for transmission over a binary symmetrical channel with feedback, *Problems Inform. Transmission,* vol. **12** (1976), 85–97, translated from *Problemi Peredachi Informatsii,* vol. **12**, no. 3 (1976), 3–19 (in Russian).

Christian Deppe
University of Bielefeld
Department of Mathematics
P.O. Box 100131, D-33501 Bielefeld
Germany

cdeppe@mathematik.uni-bielefeld.de

BOLYAI SOCIETY
MATHEMATICAL STUDIES, 16

Entropy, Search, Complexity, pp. 71–83.

Nonadaptive and Trivial Two-Stage Group Testing with Error-Correcting d^e-Disjunct Inclusion Matrices

A. G. D'YACHKOV*, A. J. MACULA† and P. A. VILENKIN*

We discuss three types of inclusion matrices (subset, subspace and sequence). We exhibit their disjunct properties and their applications to error-correcting nonadaptive group testing. Under some limited conditions, these structures are optimal for their disjunct properties.

1. Group testing

Suppose we have a finite ground set or *population* containing elements that can be uniquely characterized as positive or negative. We refer to the collection of positive elements, which is initially unknown, as the *positive subset P*. In the abstract *group testing problem*, P be identified by performing 0, 1 tests on subsets or *pools* of the population. A pool is said to be positive (1) if the test result indicates that a member of P is in that pool; the pool is said to be negative (0) if test result indicates otherwise. A deterministic pooling design algorithm is collection of pools along with a (worst case) method that identifies the positive subset in a population.

Although research in group testing has continued since Dorfman's 1943 paper [5], a renewed interest in the subject has occurred largely because of the application of group testing to bioinformatics (e.g., clone library screening [3] and ogilio array quality control [4]). See Du and Hwang's

*Supported in part by NSF DMS 0107179.
†Supported in part by Russian Foundation of Basic Research 98-01-00241.

monograph, *Combinatorial Group Testing and its Applications* [6] and its extensive bibliography.

In *nonadaptive group testing* (NGT,) one must decide exactly which pools to test *before any testing occurs.* A nonadaptive group testing algorithm is sometimes referred to as a *one-stage* algorithm. *Adaptive algorithms* are multi-stage algorithms and group testing algorithms can be classified by the degree to which they are adaptive. A *two-stage* algorithm is a *nearly* nonadaptive algorithm, in which, an initial battery of tests is carried out (often in parallel.) Then using the information from the first stage, the second and final battery of tests are constituted and carried out (again, in parallel if possible.) Because bioinformatic applications are often automated, nonadaptive or two-stage rather than adaptive methods are generally preferred here. See [13] for other factors that favor nonadaptive over adaptive methods.

The nonadaptive (or *predetermined* [2]) versions of search problems (e.g., counterfeit coins) have been studied for decades. See [1], [2] and [18]. And, equivalent formulations of NGT existed well before the advent of bioinformatics. The first equivalent formulation of NGT was introduced by Kautz and Singleton in the investigation of *binary superimposed codes* (BSC) [19] and their work was advanced by D'yachkov and Rykov [8], [9], [10]. Another equivalent formulation of NGT was studied by Erdős, Frankl and Füredi in their investigations of *cover-free families* (CFF) [11], [12]. (Also see [14], [25].) The concepts NGT, BSC, and CFF are mutually equivalent because they all have the same *disjunct* matrix representation. See Chapter 7 in [6] for references and a description. In this paper, we discuss the disjunct properties of some inclusion matrices.

2. Inclusion matrices

In this paper, all simple lower case roman variables are non-negative integers. Let $[n]$ denote $\{1, 2, \ldots, n\}$. Given set S, $|S|$ denotes its cardinality. We call a subset of $[n]$ with cardinality k a *k-set.* Let $\binom{[n]}{k}$ denote the k-sets of $[n]$. When we say binary matrix, we mean a 0, 1 matrix and vice-versa.

Suppose Γ is a family of k-sets on $[n]$ such that $|K_1 \setminus K_2| \geq r$ for all $K_1, K_2 \in \Gamma$. In the other words, the incidence vectors (of length n) of the members of Γ form a binary code with minimum Hamming distance $2r$.

Definition 1. Let $1 \leq d \leq k \leq n$ and let Γ be a family of k-sets on $[n]$ with minimum Hamming distance $2r$. We define the $\binom{n}{d} \times |\Gamma|$ binary matrix $\delta(n,d,k,\Gamma,r)$ by letting its rows and columns be respectively indexed by the members of $\binom{[n]}{d}$ and Γ. For $D \in \binom{[n]}{d}$ and $K \in \Gamma$ the matrix $\delta(n,d,k,\Gamma,r)$ has a 1 in its $(D,K)^{\text{th}}$ entry if and only if $D \subseteq K$. If $\Gamma = \binom{[n]}{k}$, then we simply write $\delta(n,d,k)$ instead of $\delta(n,d,k,\Gamma,1)$.

The general matrices $\delta(n,d,k,\Gamma,r)$ were studied in [7] and [22]. Gottlieb studied the specific matrices $\delta(n,d,k)$ in [15] from a linear algebra point of view. (Also see [20], [27] and [28].) The constant row and column weights of $\delta(n,d,k)$ are $\binom{n-d}{k-d}$ and $\binom{k}{d}$ respectively. In Figure 1, $\delta(4,2,3)$ is displayed.

$\delta(4,2,3)$	$\{1,2,3\}$	$\{1,2,4\}$	$\{1,3,4\}$	$\{2,3,4\}$
12	1	1	0	0
13	1	0	1	0
14	0	1	1	0
23	1	0	0	1
24	0	1	0	1
34	0	0	1	1

Figure 1

The weight of a vector is the number of its non-zero entries. For $k \leq n$, let $\binom{[n]}{k}[q]^k$ denote the set of all q-ary vectors of length n and weight k. Let $\alpha \in \binom{[n]}{d}[q]^d$ and $\tau \in \binom{[n]}{k}[q]^k$. Let $\alpha(i)$ and $\tau(i)$ denote the ith entry of α and τ respectively. We say that $\alpha \prec \tau$ if, for $1 \leq i \leq n$, $\alpha(i) = \tau(i)$ whenever $\alpha(i) \neq 0$. For example, $03010 \prec 03112$ where $03010 \in \binom{[5]}{2}[3]^2$ and $03112 \in \binom{[5]}{4}[3]^4$.

Definition 2. Let $1 \leq d \leq k \leq n$ and $q \geq 1$. We define the $\binom{n}{d}q^d \times \binom{n}{k}q^k$ binary matrix $\pi(q,n,d,k)$ by letting its rows and columns be respectively indexed by the members of $\binom{[n]}{s}[q]^d$ and $\binom{[n]}{k}[q]^k$. For $\alpha \in \binom{[n]}{d}[q]^d$ and $\tau \in \binom{[n]}{k}[q]^k$ the matrix $\pi(q,n,d,k)$ has a 1 in its $(\alpha,\tau)^{\text{th}}$ entry if and only if $\alpha \prec \tau$.

The row and column weights of $\pi(q,n,d,k)$ are $\binom{n-d}{k-d}q^{n-d}$ and $\binom{k}{d}$ respectively. In Figure 2, $\pi(2,4,2,3)$ is depicted.

The matrices $\pi(q,n,d,k)$ (and $\delta(n,d,k)$) are particular instances of a more general type of incidence matrices discussed in [23]. (The general construction in [23] is too lengthy for this brief survey.) It would bc interesting to see if some of the results below can be extended (in some way) to the more general setting.

	0	0	0	0	0	1	1	1	1	1		2	2
	1	1	1	2	2	0	0	0	0	1		2	2
$\delta(2,4,2,3)$	1	1	2	1	2	1	1	2	2	0	…	1	2
	1	2	2	2	2	1	2	1	2	1		0	0
0011	1	0	0	0	0	1	0	0	0	0		0	0
0012	0	1	0	1	0	0	1	0	0	0	…	0	0
0021	0	0	0	0	0	0	0	1	0	0		0	0
0022	0	0	1	0	1	0	0	0	1	0		0	0
0101	1	0	0	0	0	0	0	0	0	1		0	0
0102	0	1	1	0	0	0	0	0	0	0	…	0	0
0110	1	1	0	0	0	0	0	0	0	0		0	0
0120	0	0	1	0	0	0	0	0	0	0		0	0
⋮					⋮			⋮				⋮	
2100	0	0	0	0	0	0	0	0	0	0	…	0	0
2200	0	0	0	0	0	0	0	0	0	0		1	1

Figure 2

Let q be a prime power and let F_q^n denote the n-dimensional vector space over the finite field F_q. For $1 \le k \le n$, the *Gaussian coefficients* $\left[{n \atop k}\right]_q$ give the number of k-dimensional subspaces of F_q^n. Let $\left[{[n] \atop k}\right]_q$ denote the family of k dimensional subspaces of F_q^n. In many ways $\left[{n \atop k}\right]_q$ generalizes $\binom{n}{k}$. See [27]. It is well known that

$$\begin{bmatrix} n \\ k \end{bmatrix}_q = \prod_{i=0}^{k-1} (q^{n-i} - 1)(q^{k-i} - 1)^{-1}.$$

Definition 3. Let $1 \le d \le k \le n$ and let q be a prime power. We define the binary matrix $\gamma(n,d,k)$ by letting its the rows and columns be respectively indexed by the members of $\left[{[n] \atop d}\right]_q$ and $\left[{[n] \atop k}\right]_q$. For $D \in \left[{[n] \atop d}\right]_q$ and $K \in \left[{[n] \atop k}\right]_q$ the matrix $\gamma(n,d,k)$ has a 1 in its $(D,K)^{\text{th}}$ entry if and only if D is a subset (hence a subspace) of K.

The matrix $\gamma(n,d,k)$ was studied in [28] from a linear algebra point of view and in [24] from a group testing point of view. The constant row and column weights of $\gamma(n,d,k)$ are $\left[{n-d \atop k-d}\right]_q$ and $\left[{k \atop d}\right]_q$ respectively.

3. d^e-DISJUNCT MATRICES

Let M be an $n \times t$ binary matrix. Let $\big(c(i)\big)$ be a column (vector) of M where $1 \le i \le n$ and $c(i)$ is the ith coordinate. In many cases, we just write c in place of $\big(c(i)\big)$. Let $T = \{c_1, \ldots, c_d\}$ be a d-set of columns of M. Let c_0 be a column of M with $c_0 \notin T$. We say that (c_0, T) is a *designated* $(d+1)$*-set*. We say that a row of the matrix M *separates the designated* $(d+1)$*-set* (c_0, T) if that row has a 1 in the designated c_0 column and 0s in all the columns of S. For binary n-vectors c_1, c_2, we write $c_1 \le c_2$ if and only if $c_1(i) \le c_2(i)$ for all $1 \le i \le n$. Let S be a subset of the columns of M. Then $\vee S$ denotes the (coordinate-wise) sup of the columns in S.

Definition 4. Let $1 \le d < t$. An $n \times t$ binary matrix M is called d^e*-disjunct* if and only if given any designated $(d+1)$-set (c_0, T), there are $e+1$ rows of M that separate it.

When $e = 0$ our Definition 4 coincides with the definition of d-disjunct that is found in Chapter 7 of [6]. In that case we just say d-disjunct instead of d^0-disjunct. Obviously d^e-disjunct implies $s^{e'}$-disjunct when $d \ge s$ and $e \ge e'$.

Suppose that in a population of size t, the positive subset P has at most d elements. Then an $n \times t$ d-disjunct matrix M gives a deterministic pooling design and algorithm in the following way. Let $\big(c(i)\big)$ where $1 \le i \le n$ be a column (vector) of M. Identifying the columns of M with the population, then the rows of M give the pools in the obvious way. That is, a column $\big(c(i)\big)$ is in the pool determined by the r_i (the row of M with index i) if and only if the (column) entry $c(i) = 1$. The information gained by testing these pools is organized as follows. Suppose that the positive subset is $P = \big\{\big(c_j(i)\big)\big\}_{j \in S}$. By testing each pool (row) r_i, we define an *output vector* $\big(o(i)\big)$ by setting $o(i) = 1$ if pool r_i is positive and $o(i) = 0$ if it is negative. Clearly (given that the tests are error-free) for $1 \le i \le n$, $o(i) = 1$ if and only if there is a $\big(c_j(i)\big) \in P$ with $c_j(i) = 1$. Thus $o = \vee P$. The output vector o is used to identify P because $P = \{c \in M : c \le o\}$. This follows because for each $c_0 \notin P$ there is a row of M that separates the designated set (c_0, P).

In a somewhat different way, a d^e-disjunct matrix can be used to identify P even if some errors occur in the output vector. Note that in the above error-free situation the positive subset P could have been identified by checking all subsets S of columns of M with cardinality at most d for the only such S with $\vee S = o$. This is because a d-disjunct matrix has the property (strictly weaker that d-disjunct) that all such $\vee S$ are distinct. So if $\vee S = o$, then it follows that $S = P$. Matrices with this weaker property are called $\bar{d}$-separable. See [6]. What follows below is inspired by [16] and [24].

Proposition 1. *Let $1 \leq d < t$ and let M be an $n \times t$ d^e-disjunct matrix. Let $H(x, y)$ be the Hamming distance between two binary vectors x and y of length n. Let S and T be distinct subsets of columns of M with cardinality at most d. Then:*

a. [16] *If $S \subset T$, then $H(\vee S, \vee T) \geq e + 1$.*

b. *If $S \not\subset T$ and $T \not\subset S$, then $H(\vee S, \vee T) \geq 2e + 2$.*

Proof. a. Since there is a column $c_0 \in T$ with $c_0 \notin S$, then the designated set (c_0, S) is separated by at least $e + 1$ rows of M. From this the result follows.

b. Since there is a column $c_0 \in T$ with $c_0 \notin S$ and a column $c_1 \in S$ with $c_1 \notin T$, then the designated sets (c_0, S) and (c_1, T) are each separated by at least $e + 1$ rows of M. From this the result follows. ■

Thus depending upon the prior knowledge about the nature of the positive subset P, it follows from Proposition 1, that d^e-disjunct matrices can be use to correct a varying number of errors.

In the most general situation where the assumption is that the cardinality of P is at most d, then a d^e-disjunct can correct $\lfloor e/2 \rfloor$ errors in the following way. Suppose that at most $\lfloor e/2 \rfloor$ errors have occurred in the correct output o and, in place of o, ε is the *observed error output vector.* Then $H(o, \varepsilon) = H(\vee P, \varepsilon) \leq \lfloor e/2 \rfloor$. By checking all subsets S of columns of M with cardinality *at most* d for the only one with $H(\vee S, \varepsilon) \leq \lfloor e/2 \rfloor$, then $S = P$. Clearly there is at least one such S (e.g., $S = P$) and if $S \neq P$, then, by Proposition 1a, we have that $H(\vee S, \varepsilon) + H(\varepsilon, o) \geq H(\vee S, o) = H(\vee S, \vee P) \geq e+1$. Since $H(e, o) \leq \lfloor e/2 \rfloor$, it follows that $H(\vee S, \varepsilon) > \lfloor e/2 \rfloor$.

If we assume that P has cardinality exactly d, then a d^e-disjunct matrix can correct e errors in a similar way. Suppose that at most e errors have occurred in the correct output o and δ is the observed (error) output vector. In this case, $H(o, \delta) = H(\vee P, \delta) \leq e$. By checking all subsets S of columns of M with cardinality *exactly* d for the only one with $H(\vee S, \delta) \leq e$, then $S = P$. This follows from Proposition 1b because if $S \neq P$, then $H(\vee S, \delta) + H(\delta, o) \geq H(\vee S, o) = H(\vee S, \vee P) \geq 2e + 2$. Since $H(\delta, o) \leq e$, it follows that $H(\vee S, \delta) > e$.

One question is that if we take the more general assumption that the cardinality of P is at most d, can a d^e-disjunct matrix correct more that $\lfloor e/2 \rfloor$ errors? The answer to this question is "yes and no". The answer is "no" if a nonadaptive procedure is desired. The following counter-example from [16] demonstrates this.

Example 1. The matrix in Figure 3a is d^e-disjunct with $d = 2$ and $e = 1$. If the positive subset is the first two columns, then the correct output vector is o_1 in Figure 3b. If the positive subset is just the first column, then the correct output vector is o_2 in Figure 3b. Since both of these are Hamming distance one away from an error output vector in δ in Figure 3c, then with one error in the testing, it is impossible to determine a unique positive subset.

$$\begin{pmatrix} 1 & 0 & 0 \\ 1 & 0 & 0 \\ 0 & 1 & 0 \\ 0 & 1 & 0 \\ 0 & 0 & 1 \\ 0 & 0 & 1 \end{pmatrix} \qquad \mathbf{o}_1 = \begin{pmatrix} 1 \\ 1 \\ 1 \\ 1 \\ 0 \\ 0 \end{pmatrix} \qquad \mathbf{o}_2 = \begin{pmatrix} 1 \\ 1 \\ 0 \\ 0 \\ 0 \\ 0 \end{pmatrix} \qquad \delta = \begin{pmatrix} 1 \\ 1 \\ 1 \\ 0 \\ 0 \\ 0 \end{pmatrix}$$

$(a) \qquad\qquad (b) \qquad\qquad (c) \qquad\qquad (d)$

Figure 3

While Example 1 shows that it is hopeless to nonadaptively use a d^e-disjunct matrix to correct e testing errors when it is assumed that the cardinality of P is at most d, it also indicates that a d^e-disjunct matrix can be used to correct e testing errors when it is assumed that the cardinality of P is at most d and a *trivial two-stage* decoding method is used. A trivial two-stage decoding method adds a second stage that tests each member of a relatively small group of the population individually and each of these individual tests are guaranteed to be error free. We describe this below.

Assume that the positive subset P has cardinality at most d. Suppose that at most e errors have occurred in the correct output o and, in place of o, δ is the observed error output vector. Take all subsets S of columns of M with cardinality *at most* d and with $H(\vee S, \delta) \leq e$. Let S_1 and S_2 be two such subsets. Then either $S_1 \subset S_2$ or $S_2 \subset S_1$. If not, then by Proposition 1b, we have $H(\vee S_1, \vee S_2) \geq 2e + 2$. Then either $H(\vee S_1, \delta) > e$ or $H(\vee S_2, \delta) > e$ which is a contradiction. Thus set of all at most d sets S with $H(\vee S, \delta) \leq e$ is a chain. Since $H(\vee P, \delta) = H(o, \delta) \leq e$, then P

is a member of this chain. If one simply takes the union of all sets S of cardinality at most d with, then this union can have cardinality at most d. Since P is a member of this union, it can be identified by testing each member of the union individually. The main point is that one doubles the error-correcting capabilities by the addition of at most d *confirmatory and guaranteed* tests as compared to the number of tests required by, and error correcting capabilities of, the purely nonadaptive case.

Proposition 2. *Let $1 \le s \le d \le k \le n$. Let $1 \le q$ and $e = \binom{k-s}{d-s} - 1$.*

a. [7], [21] *$\delta(n,d,k)$ is s^e-disjunct.*

b. [23] *$\pi(q,n,d,k)$ is s^e-disjunct.*

Proof. We give a new proof of part a. The proof of part b is similar. First, it follows from induction that

$$\sum_{i=0}^{s} (-1)^i \binom{s}{i}\binom{k-i}{d} = \binom{k-s}{d-s}. \tag{1}$$

Let $S = \{K_0, K_1, \ldots, K_s\}$ be the (set of) k-set indices of $s+1$ columns of $\delta(n,d,k)$. Suppose the column with index K_0 is designated. It suffices to show that there are at least $\binom{k-s}{d-s}$ d-sets in K_0, none of which are contained in any K_i with $1 \le i \le s$, because then we would have $\binom{k-s}{d-s}$ rows of $\delta(n,d,k)$ separating (K_0, S). Without loss of generality, we can assume that $|K_0 \cap K_i| = k-1$ for all $1 \le i \le s$, because such a family $\{K_1, \ldots, K_s\}$ will cover the most possible number of d-sets of K_0. However, the number of d-sets of K_0 that are *not* covered by some member of such a family $\{K_1, \ldots, K_s\}$ is given by (1) because the intersection of any $i(k-1)$-sets of $[k]$ has cardinality $k-i$. ■

Proposition 3. *For $1 \le d \le k \le n$ and $1 \le r < k$, let Γ be a family of k-sets in $[n]$ with minimum Hamming distance $2r$. Let*

$$p = \left\lceil \left(\binom{k}{d} - \binom{k-r}{d}\right)\left(\binom{k-r}{d} - \binom{k-2r}{d}\right)^{-1} \right\rceil$$

and for $1 \le s \le p$, let

$$e = \binom{k}{d} - \binom{k-r}{d} - (s-1)\left(\binom{k-r}{d} - \binom{k-2r}{d}\right) - 1.$$

Then for $1 \le s \le p$, we have that $\delta(n,d,k,\Gamma,r)$ is s^e-disjunct.

Proof. $S = \{K_0, K_1, \ldots, K_s\}$ be the (set of) k-set indices of $s+1$ columns of $\delta(n, d, k, \Gamma, r)$. Suppose the column with index K_0 is designated. It suffices to show that there are at least $e+1$ (as given above) d-sets in K_0 none of which are contained in any K_i with $1 \leq i \leq s$. Without loss of generality, we can assume that $|K_0 \cap K_i| = k-r$ for all $1 \leq i \leq s$, because such a family $\{K_1, \ldots, K_s\}$ will cover the most possible number of d-sets of K_0. K_1 covers $\binom{k-r}{d}$ d-sets of K_0. Let $1 \leq i \leq s-1$. Since $|K_{i+1} \cap K_i \cap K_0| \geq k - 2r$, then the number of d-sets of K_0 covered by K_{i+1} *and* not already covered by K_i is $\binom{k-r}{d} - \binom{k-2r}{d}$. From this the result follows. ■

Proposition 3 here improves Proposition 4 in [7].

Proposition 4. *Let q be a prime power. For $1 \leq d \leq k \leq n$, let*

$$p = \left\lceil \left(\begin{bmatrix} k \\ d \end{bmatrix}_q - \begin{bmatrix} k-1 \\ d \end{bmatrix}_q \right) \left(\begin{bmatrix} k-1 \\ d \end{bmatrix}_q - \begin{bmatrix} k-2 \\ d \end{bmatrix}_q \right)^{-1} \right\rceil .$$

For $1 \leq s \leq p$, let

$$e = \begin{bmatrix} k \\ d \end{bmatrix}_q - \begin{bmatrix} k-1 \\ d \end{bmatrix}_q - (s-1) \left(\begin{bmatrix} k-1 \\ d \end{bmatrix}_q - \begin{bmatrix} k-2 \\ d \end{bmatrix}_q \right) - 1.$$

Then $1 \leq s \leq p$, we have that $\gamma(n, d, k)$ is s^e-disjunct.

Proof. Let $S = \{K_0, K_1, \ldots, K_s\}$ be the k-space indices of $s+1$ columns of $\gamma(n, d, k)$. Suppose the column with index K_0 is designated. It suffices to show that there are at least $e+1$ (as given above) d-spaces in K_0 none of which are contained in any K_i with $1 \leq i \leq s$. Without loss of generality, we can assume that the dimension of $K_i \cap K_0$ is $k-1$ for all $1 \leq i \leq s$, because such a family $\{K_1, \ldots, K_s\}$ will cover the most possible number of d-spaces of K_0. K_1 covers $\begin{bmatrix} k-1 \\ d \end{bmatrix}_q$ d-spaces of K_0. Let $1 \leq i \leq s-1$. Since the intersection of any two $(k-1)$-spaces in a k-space is a $(k-2)$-space, then number of d-spaces of K_0 covered by K_{i+1} *and* not already covered by K_i is $\begin{bmatrix} k-1 \\ d \end{bmatrix}_q - \begin{bmatrix} k-2 \\ d \end{bmatrix}_q$. From this the result follows. ■

Proposition 4 is an improvement of a Theorem 3.4 in [24]. The following Corollary 1 is a special case of Proposition 4.

Corollary 1. *Let q be a prime power. For $1 \leq s \leq q$, let*

$$e = q^{k-1} - (s-1)q^{k-2} - 1.$$

Then for $1 \leq s \leq q$, we have that $\gamma(n, 1, k)$ is s^e-disjunct.

4. Some optimal d^e-disjunct matrices with constant row weight

In applications of group testing, there are limits on the size of any given pool. The size of pool in a deterministic NGT method is given by the row weight of the matrix representation of the pooling design. Proposition 5 is a reformulation of Proposition 1 in [7].

Proposition 5 [7]. *Let M be an $n \times t$ d^e-disjunct matrix. Let r be the maximum row weight of M. Suppose that $(e+1)r \geq d+e+1$ and $n \geq 2$. Then $n \geq \left\lceil \frac{t(d+e+1)}{r} \right\rceil$.*

Proposition 6. *Suppose $d \leq t-r$. Let M be an $n \times t$ d^e-disjunct matrix with constant row weight r. Then*

$$n \geq \left\lceil (d+1)(e+1)\binom{t}{d+1}\left(r\binom{t-r}{d}\right)^{-1} \right\rceil .$$

Proof. There are $(d+1)\binom{t}{d+1}$ designated $(d+1)$-sets of columns of M and each needs to be separated by $e+1$ rows. Since each row separates exactly $r\binom{t-r}{d}$ designated $(d+1)$-sets the result follows. ■

Sometimes the bound in Proposition 6 exceeds that in Proposition 5. If a d^e-disjunct matrix with constant row weight achieves either bound, we say that it is *optimal for its (maximum, constant) row weight.* Part a of Corollary 2 comes from [7].

Corollary 2. *Let $1 \leq s \leq d < n$ and $e = \binom{k-s}{d-s} - 1$. Then*

a. [7] *$\delta(n,d,d+1)$ is s^e-disjunct and it is optimal for its constant row weight of $n-d$.*

b. *$\pi(q,n,d,d+1)$ is s^e-disjunct and it is optimal for its row constant row weight of $q(n-d)q$.*

Proof. Apply Proposition 2. It is straightforward to verify that the bound in Proposition 5 is achieved. ■

Corollary 3. *For $1 \leq d < n$ and let q be a prime power. Then for $1 \leq s \leq q$, we have $\gamma(n,d,d+1)$ is s^e-disjunct with $e = q-s$ and it is optimal for its constant row weight of $\begin{bmatrix} n-d \\ 1 \end{bmatrix}_q = (q^{n-d}-1)(q-1)^{-1}$.*

Proof. Since

$$\left\lceil \begin{bmatrix} d+1 \\ d \end{bmatrix}_q - \begin{bmatrix} d \\ d \end{bmatrix}_q \right\rceil = q,$$

then apply Proposition 4. It is straightforward to verify that the bound in Proposition 5 is achieved. ■

5. Remarks

Recall that d^e-disjunct implies $s^{e'}$-disjunct when $d \geq s$and $e \geq e'$. And an $s^{e'}$-disjunct matrix *could* also be d^e-disjunct with $d \geq s$ and/or $e \geq e'$. We say that an $s^{e'}$-disjunct matrix M is *fully* if it is not d^e-disjunct whenever $d \geq s$ or $e \geq e'$. Proposition 2 and Corollary 1 can not be "improved" because those statements remain true if we substitute the phrase "fully s^e-disjunct" in place of "s^e-disjunct". However, Proposition 3 and Proposition 4 with $d \neq 1$ could possibly be improved in so much as it is unlikely that they would remain true if the phrase "fully s^e-disjunct" was substituted for "s^e-disjunct" therein.

References

[1] R. Ahlswede and I. Wegener, *Search Problems,* Wiley, New York (1987).

[2] M. Aigner, *Combinatorial Search,* Teubner, Stuttgart (1988).

[3] D. J. Balding, et al., A comparative survey of non adaptive pooling designs, Genetic Mapping and DNA Sequencing, in: *IMA Volumes in Mathematics and its Applications,* Springer Verlag (1995), 133–155.

[4] C. Colburn, A. Ling and M. Tompa, Construction of optimal quality control ogilio arrays, *Bioinformatics,* **18**, no. 4 (2002), 529–535.

[5] R. Dorfman, The detection of defective members of a large population, *Ann. Math. Stat.,* **14** (1943), 436–440.

[6] D-Z. Du and F. Hwang, *Combinatorial Group Testing and Its Applications,* 2nd ed. World Scientific, Singapore (2000).

[7] A. Dyachkov, A. Macula and V. Rykov, New applications and results of superimposed code theory arising from the potentialities of molecular biology, in: *Numbers and Combinatorics, Proc., "Numbers, Information, and Complexity",* Conference, Bielefeld Germany, Ingo Althofer, et al. eds., Kluwer Academic Publishers (2000), 265–282.

[8] A. Dyachkov and V. Rykov, Bounds on the lengths of disjunct codes, *Problems Contr. and Inf. Theory,* **11** (1982), 7–13.

[9] A. Dyachkov and V. Rykov, A survey of superimposed code theory, *Problems Contr. and Inf. Theory,* **12** (1983), 1–13.

[10] A. Dyachkov and V. Rykov, Superimposed distance codes, *Problems Contr. and Inf. Theory,* **18** (1989), 237–250.

[11] P. Erdős, P. Frankl and Z. Füredi, Families of finite sets in which no set is covered by the union of two others, *Jour. Comb. Th.,* **A33** (1982), 158–166.

[12] P. Erdős, P. Frankl and Z. Füredi, Families of finite sets in which no set is covered by the union of r others, *Israel J. Math.,* **51** (1985), 79–89.

[13] Farach, et al., Group testing problems with sequences experimental molecular biology, in: *Proceedings of Compression and Complexity of Sequences,* 1997, B. Carpentieri, et al. (eds.) IEEE Press (1997), 357–367.

[14] Z. Füredi, On r-cover-free families, *Jour. Combin. Theory Ser. A,* **73** (1996), 172–173.

[15] D. Gottlieb, A certain Class of incidence matrices, *Proc. Amer. Math. Soc.,* **17** (1966), 1233–1237.

[16] T. Huang and C. Weng, Pooling spaces and non-adaptive pooling designs, *Discrete Math.,* **282** (2004), 163–169.

[17] F. Hwang and V. Sós, Non-adaptive hypergeometric group testing, *Studia Sci. Math. Hung.,* **22** (1987), 257–263.

[18] G. Katona, Rényi and the combinatorial search problems, *Studia Sci. Math. Hung.,* **26** (1991), 363–376.

[19] W. Kautz and R. Singleton, Nonrandom binary superimposed codes, *IEEE Trans. Inf. Theory,* **10** (1964), 363–377.

[20] G. Khosrovshahi and Ch. Maysoori, On the structure of higher incidence matrices, *Bull. Inst. Combin. Appl.,* **25** (1999), 13–22.

[21] A. Macula, A simple construction of d-disjunct matrices with certain constant weights, *Discrete Mathematics,* **162** (1996), 311–312.

[22] A. Macula, Nonadaptive group testing with error-resistant d-disjunct matrices, *Discrete Applied Mathematics,* **80** (1998), 217–282.

[23] A. Macula and P. Vilenkin, Constructions of superimposed codes based on incidence structures, with P. A. Vilenkin, in: *2000 IEEE Proceedings of International Symposium on Information Theory* (2000), p. 3.

[24] H., Ngo and D-Z. Du, New constructions of non-adaptive and error-tolerance pooling designs, *Discrete Math.,* **243** (2002), 161–170.

[25] M. Ruszinkó, On the upper bound of the size of r-cover free families, *J. Comb. Th. A,* **66** (1994), 302–310.

[26] J. van Lint and R. Wilson, *A Course in Combinatorics,* Cambridge Univ. Press, Cambridge (1992).

[27] R. Wilson, A diagonal form for the incidence matrices of t-subsets vs. k-subsets, *European J. Combin.*, **11**, no. 6 (1990), 609–615.

[28] A. Yakir, Inclusion martrix of k vs. l affine subspaces and a permutation module of the general affine group, *J. Combin. Theory Ser. A*, **63**, no. 2 (1993), 301–317.

Arkadii G. D'yachkov and
Pavel A. Vilenkin
Department of Probability Theory
Faculty of Mechanic and Mathematics
Moscow State University
Moscow, Russia 119899
dyachkov@mech.math.msu.su
paul@vilenkin.dnttm.ru

Anthony J. Macula
Department of Mathematics
SUNY Geneseo
Geneseo, NY 14454
macula@geneseo.edu

BOLYAI SOCIETY MATHEMATICAL STUDIES, 16
Entropy, Search, Complexity, pp. 85–112.

Model Identification Using Search Linear Models and Search Designs

S. GHOSH, T. SHIRAKURA and J. N. SRIVASTAVA

1. Introduction

In designing an experiment, we often assume a model and then find a *best* design satisfying one or more *optimal* properties under the assumed model. This approach works well when we are absolutely sure about the assumed model that it will fit the experimental data adequately. In reality, we are rarely sure about a particular model in terms of its effectiveness in describing the data adequately. However, we are normally sure about a set of possible models that would describe the data adequately and one of them would possibly describe the data better than the other models in the set. The pioneering work of Srivastava [33] introduced the search linear model with the purpose of searching for and identifying the correct model from a set of possible models that includes the correct model. This paper focuses on model identification through the use of the search linear models, particularly in addressing the fundamental issues and important challenges in statistical design and analysis of experiments while presenting an overview on this area of research. Two important research areas developed over time using the search linear models are in factorial designs and row-column designs.

In factorial experiments, we normally assume that the lower order effects are important and the higher order effects are all negligible. For example, in a main-effect plan, we assume that the factors do not interact, or in other words, the interaction effects are all negligible. Such an assumption may or may not be true in reality because of a possible presence of a few significant non-negligible effects. The standard linear models cannot identify these non-negligible effects using a small number of runs or treatments considerably smaller than the total number of possible runs for an experiment. This motivates the use of search designs under the search linear model in searching for and identifying the non-negligible effects.

In row-column designs there are two blocking factors, namely the row-blocking factor and the column-blocking factor. Since the goal of an experiment is to compare treatments and their effects, such blocking factors are considered as nuisance factors. In the standard analysis of row-column designs, the two nuisance factors are assumed to be additive meaning that they do not interact and there are no interaction effects. Such nuisance factors arising in nature will not always be additive. Srivastava [34, 37, 38], Srivastava and Beaver [40] considered such non-additive nuisance factors leading to nested multidimensional block designs. Srivastava and Wang [46] developed a technique for examining possible non-additive effects and identifying the non-additive cells in row-column designs using the search linear models. Some experimental strategies were also developed to eliminate the influence of non-additive effects on the analysis of row-column designs.

Srivastava [35] introduced the concepts *sensitivity* and *revealing power* in experimental designs. Sensitivity as a generalization of the concept of local control, deals with the determination of grouping units so that the effects of known and possibly even unknown nuisance factors can be eliminated as much as possible. Revealing power refers to the ability of the design to find out what the true model might be in a particular experimental situation.

This paper has eight sections. Sections 1–3 are written by S. Ghosh, 4 by T. Shirakura, and 5–8 by J. N. Srivastava. In Sections 2 and 3, Ghosh describes the search linear models, search procedure, search probabilities, and performance evaluation of the search procedure and search designs. In Section 4, Shirakura presents the details about search designs in factorial experiments. In Sections 5–8, Srivastava addresses the fundamental issues in scientific experiments on model identification, examines the information contained in an experiment about the model, presents some combinatorial problems, and describes the analysis under the general search linear model using simulation distributions.

2. Search linear model

Consider the search linear model (Srivastava [33])

$$(1) \qquad E(\mathbf{y}) = \mathbf{A}_1\xi_1 + \mathbf{A}_2\xi_2, \quad V(\mathbf{y}) = \sigma^2\mathbf{I},$$

where $\mathbf{y}(n \times 1)$ is the vector of observations, $\mathbf{A}_1(n \times \nu_1)$, and $\mathbf{A}_2(n \times \nu_2)$ are matrices known from the underlying design. The elements of the vector $\xi_1(\nu_1 \times 1)$ are unknown parameters. About the elements of $\xi_2(\nu_2 \times 1)$, we know that at most k elements are nonzero but we do not know which elements are nonzero. The goal is to search for and identify the nonzero elements of ξ_2 and then estimate them along with the elements of ξ_1. Such a model is called a search linear model (SLM). When $\xi_2 = \mathbf{0}$, the search linear model becomes the ordinary linear model. For the search linear model, we have $\xi_2 \neq \mathbf{0}$.

Let $\mathbf{A}_{22}$ be any $(n \times 2k)$ submatrix obtained from $\mathbf{A}_2$. A design is a search design (Srivastava [33]) if, for every submatrix $\mathbf{A}_{22}$,

$$(2) \qquad \text{Rank}\,[\mathbf{A}_1, \mathbf{A}_{22}] = \nu_1 + 2k.$$

The rank condition (2) allows us to fit and discriminate between any two models in the class of possible models described earlier. Thus, a search design allows us to search for and identify the nonzero elements of ξ_2 and then estimate them along with the elements of ξ_1.

When $\text{Rank}\,[\mathbf{A}_1] = \nu_1$, the condition (2) is equivalent (Srivastava and Ghosh [42]) to

$$(3) \qquad \text{Rank}\left[\mathbf{A}_{22}'\mathbf{A}_{22} - \mathbf{A}_{22}'(\mathbf{A}_1'\mathbf{A}_1)^{-1}\mathbf{A}_{22}\right] = 2k.$$

3. Search procedure, search probabilities, and performance evaluation

Consider a class of $\binom{\nu_2}{k}$ linear models from (1) with the parameters as ξ_1 and k elements of ξ_2. The $\binom{\nu_2}{k}$ possible sets of k elements of ξ_2 give rise to $\binom{\nu_2}{k}$ such models. For any two models in this class, the elements ξ_1 are common parameters but in the two sets of k elements in ξ_2 some common parameters may or may not be present. A search procedure identifies a model which best fits the data generated from the search design(SD). To identify such a model, the sum of squares of error (SSE) of each model is used (Srivastava [33]). If SSE for the first model ($M1$) is smaller than the SSE for the second model ($M2$), then $M1$ provides a better fit and is selected over $M2$. For a fixed value of k, all $\binom{\nu_2}{k}$ models are fitted to the data and the search procedure selects the model with the smallest SSE as the best model for describing the data.

Two popular criteria for model comparisons, namely the Akaike Information Criterion (AIC) and the Bayesian Information Criterion (BIC) (Akaike [1, 2], Sawa [21]) turn out to be exactly equivalent to the minimization of SSE in the above search procedure.

The probability of selecting a model over another model depends on σ^2. To explain this, let $M0$ be the true model in the class of models described above and $M1$ be a competing model where $M1 \neq M0$. In the noiseless case, $\sigma^2 = 0$, and the SSE for $M0$, SSE$(M0)$, is zero, which is always smaller than the SSE$(M1)$. Hence, $M0$ will definitely be selected over $M1$. Thus $P\big[\text{SSE}(M0) < \text{SSE}(M1) \mid M0, M1, \sigma^2 = 0\big] = 1$. In reality $\sigma^2 > 0$ and the SSE$(M0)$ may not be less than SSE$(M1)$. Therefore, $M0$ may not necessarily be selected over $M1$. Hence, the probability of correctly identifying the nonzero interaction is less than one, i.e., $P\big[\text{SSE}(M0) < \text{SSE}(M1) \mid M0, M1, \sigma^2 > 0\big] < 1$. In the case of infinite noise, $M0$ and $M1$ are indistinguishable, and so the probability of selecting $M0$ over $M1$ is $1/2$, i.e., $P\big[\text{SSE}(M0) < \text{SSE}(M1) \mid M0, M1, \sigma^2 = \infty\big] = 1/2$. For $0 < \sigma^2 < \infty$, $P\big[\text{SSE}(M0) < \text{SSE}(M1) \mid M0, M1, \sigma^2\big]$ is called the *search probability* for a given $M0$, $M1$, and σ^2. Note that the search probability is between $1/2$ and 1.

Shirakura, Takahashi, and Srivastava [26] presented an exact expression of the search probability under the normality assumption. This expression is also given in the equation (6) of this paper. A criterion for comparing SDs based on the search probabilities is also given in Shirakura, Takahashi, and Srivastava [26].

Ghosh and Teschmacher [14] defined a search probability matrix (SPM) whose columns represent the possible true models and rows represent the possible competing models. The off-diagonal elements of the SPM represent the search probabilities corresponding to all possible pairs of $M0$ and $M1$ for a given σ^2. Since the true model $M0$ is different from the competing model $M1$, the diagonal elements of the SPM are not meaningful and therefore are left blank. Ghosh and Teschmacher [14] used such SPMs in comparing SDs and presented three criteria for comparing SDs. One of these three criteria is the criterion given in Shirakura, Takahashi, and Srivastava [26] and other two are new. Such comparisons using the three criterion depend on an unknown parameter ρ which is essentially a *Signal-to-Noise Ratio.* Ghosh and Teschmacher [14] presented a *majority rule* for comparing designs using the search probabilities for all values of ρ.

Orthogonal array designs satisfy many optimal properties under a specified factorial experiment model. Balanced arrays and orthogonal arrays are commonly used search designs. As a consequence of comparison of SDs in Ghosh and Teschmacher [14], it is observed that the balanced array is more likely to identify the nonzero interaction than the orthogonal array obtained from the Plackett-Burman design.

For comparing SDs, Srivastava [34] introduced the six criterion of minimizing the arithmetic means and the geometric means of determinants, trace, and maximum characteristic roots of the information matrix of the $\binom{\nu_2}{k}$ possible models. Shirakura and Onishi [24] presented optimal SDs for 2^m factorials using the arithmetic mean of determinants (AD-optimality) criterion. Ghosh and Burns [11] presented four general classes of search designs for 3^m factorials in factor screeing setup with $k = 1, 2$. Two of these four classes of SDs performed better than the others under all six criterion mentioned above. Ghosh [10] presented a list of known SDs obtained by various researchers.

4. Search designs

We now consider a fractional factorial design **T** for a 2^m factorial experiment with m factors each at two levels. The design **T** can be expressed as a $(0, 1)$ matrix of $N \times m$ whose rows are treatments. Srivastava and Ghosh [42, 43] obtained search designs (SDs) for $k = 1$ satisfying the condition (3). In their setup ξ_1 is a vector consisting of the general mean, main effects and two-factor interactions with $\nu_1 = 1 + m + m(m-1)/2$ parameters, and ξ_2 is a vector of all the remaining three-factor and higher order interaction effects with $\nu_2 = 2^m - \nu_1$ parameters. Srivastava and Ghosh [42] gave SDs $(4 \leq m \leq 8)$ of N treatments satisfying $(m = 4;\ 13 \leq N \leq 15)$, $(m = 5;\ 18 \leq N \leq 31)$, $(m = 6;\ 24 \leq N \leq 40)$, $(m = 7;\ 31 \leq N \leq 68)$ and $(m = 8;\ 39 \leq N \leq 59)$. Srivastava and Ghosh [42] presented SDs with the possible minimal values of N for the range of the values of m considered.

Srivastava and Gupta [45] considered the setup where ξ_1 is a vector of the general mean and main effects with $\nu_1 = 1 + m$ parameters, and ξ_2 is a vector of all remaining effects of two-factor and higher order interactions $\nu_2 = 2^m - \nu_1$ parameters. More work in this direction for different values of m can be found in Gupta [15] and Gupta and Carvajal [16].

Srivastava [34] and Ghosh [7] investigated the problem of finding SDs for $k = 1$ when $m = 2^h - 1$ with $(h \geq 2)$. A special structure of the design $\mathbf{T}$ is considered for such investigation. Let $\mathbf{T}_1$ be a $(0, 1)$ matrix of size $2^h \times m$ such that the first h columns have all distinct row vectors, and the remaining columns are obtained from them by all possible Hadamard product, where $1 \times 1 = 0 \times 0 = 1$ and $1 \times 0 = 0 \times 1 = 0$ for the product of two elements. Ghosh [7] characterized and constructed a plan $\mathbf{T}_2$ with N_2 treatments so that the design $\mathbf{T} = \mathbf{T}_1 + \mathbf{T}_2$ with $m+1+N_2$ treatments is an SD for $k = 1$. By "+", we mean that $\mathbf{T}$ is the $(N_1 + N_2) \times m$ matrix composed of rows of $\mathbf{T}_1$ and $\mathbf{T}_2$. For the same parameter vectors ξ_1 and ξ_2, Srivastava and Gupta [45], Gupta and Carvajal [16] considered $\mathbf{T}_1 = \Omega(m, 1) + \Omega(m, m)$ and $\mathbf{T}_1 = \Omega(m, 0) + \Omega(m, 1) + \Omega(m, m)$ with $N_1 = m + 1$ and $N_1 = m + 2$ treatments, respectively, and they presented necessary conditions for plans $\mathbf{T}_2$ so that the designs $\mathbf{T} = \mathbf{T}_1 + \mathbf{T}_2$ are SDs for $k = 1$. Here, $\Omega(m, j)$ is the $\binom{m}{j} \times m$ matrix whose rows have exactly j 1-elements (weight j).

For the same setting for $\mathbf{T}_1$ as in Ghosh [7], Shirakura [22] showed that for $m = 7$ and $k = 1$, the minimum number of the N_2 in $\mathbf{T}_2$ is 7. Using properties of a BIB design for $\mathbf{T}_2$, Shirakura [22] furthermore constructed an SD for $m = 2^h - 1$ and $k = 2$ under the assumption for ξ_2 to be only of the two-factor interactions. Let ξ_1 be a vector of the general mean and main effects with $(\nu_1 = 1 + m)$ parameters, and ξ_2 be a vector of the two- and three-factor interactions $\big(\nu_2 = m(m-1)/2 + m(m-1)(m-2)/6\big)$. Suppose $\mathbf{T}_1 = \Omega(m, 0) + \Omega(m, m-1) + \Omega(m, m)$ with $N_1 = m+2$ treatments. Ghosh [8] characterized and constructed a plan $\mathbf{T}_2$ with N_2 treatments so that the design $\mathbf{T} = \mathbf{T}_1 + \mathbf{T}_2$ with $N_1 + N_2$ treatments is an SD for $k = 1$ and $3 \leq m \leq 8$. However some results in Ghosh [8] are incorrect. Ohnishi and Shirakura [19] corrected his results and gave SDs for which the numbers of treatments are not greater than those in his results except for $m = 4$.

Let ξ_1 be a vector of the general mean and main effects $(\nu_1 = 1+m)$, and ξ_2 be only of the two-factor interactions $\big(\nu_2 = m(m-1)/2\big)$. Suppose $\mathbf{T}_1 = \Omega(m, 0) + \Omega(m, 1)$ with $N_1 = m+1$ treatments, and let $\mathbf{T}_2 = [\mathbf{t}_1, \mathbf{t}_2, \cdots, \mathbf{t}_m]$, where $\mathbf{t}_j$'s are $(0, 1)$ vectors of $N_2 \times 1$. Then, Shirakura [23] showed that a design $\mathbf{T} = \mathbf{T}_1 + \mathbf{T}_2$ with $N_1 + N_2$ treatments is an SD for $k = 1$ if and only if $\mathbf{T}_2$ is such that $\mathbf{t}_i * \mathbf{t}_j$ (Hadamard product) are nonzero and distinct vectors for all $1 \leq i < j \leq m$. Using this condition, Shirakura [23] presented $\mathbf{T}_2$ with minimum values N_2^* of N_2 for $3 \leq m \leq 10$, that is, $N_2^* = 3, 3, 4, 5, 6, 6, 7, 7$ for $m = 3, \cdots, 10$, respectively.

As an example, for $m = 8$ and $N_2^* = 6$, the matrix $\mathbf{T}_2$ is given by

$$\begin{bmatrix} 1 & 0 & 1 & 1 & 1 & 0 & 1 & 0 \\ 1 & 0 & 0 & 1 & 1 & 1 & 0 & 1 \\ 1 & 1 & 0 & 0 & 1 & 1 & 1 & 0 \\ 1 & 0 & 1 & 0 & 0 & 1 & 1 & 1 \\ 1 & 1 & 0 & 1 & 0 & 0 & 1 & 1 \\ 1 & 1 & 1 & 0 & 1 & 0 & 0 & 1 \end{bmatrix}.$$

As an open problem, we are interested in finding the minimum values of N_2^* for $m \geq 11$, and in constructing $\mathbf{T}_2$ with the minimum treatments.

For the same parameter vectors ξ_1 and ξ_2 as in Ghosh [7], Srivastava and Arora [39] showed that for $m \geq 3$,

$$\mathbf{T} = \Omega(m, 0) + \Omega(m, 1) + \Omega(m, 2) + \Omega(m, m),$$

is an SD for $k = 2$ with $N = (m^2 + m + 4)/2$ treatments. In general, let ξ_1 be of up to the ℓ-factor interactions and ξ_2 be only of the $(\ell + 1)$- factor interactions, where $\nu_1 = 1 + m + \cdots + \binom{m}{\ell}$, $\nu_2 = \binom{m}{\ell+1}$ and $2(\ell + 1) \leq m$. Further consider a BA of strength $2(\ell+1)$, size N, m constraints and indices $\mu_i \ \big(i = 0, \cdots, 2(\ell + 1)\big)$ as a design $\mathbf{T}$. Then Shirakura and Ohnishi [24] presented an explicit expression for every elements of the matrix in (3) under the assumption of rank $(\mathbf{\Lambda}_1) = \nu_1$. By these results, Shirakura and Tazawa [28, 29] constructed directly SDs for $(\ell = 1; k = 1, 2)$ and $(\ell = 2; k = 1, 2)$, respectively. That is, when $\ell = 1$, Shirakura and Tazawa [28] showed that

$$\mathbf{T} = \Omega(4, 0) + \Omega(4, 1) + \Omega(4, 3), \quad m = 4,$$
$$\mathbf{T} = \Omega(m, 1) + \Omega(m, m - 1), \quad m \geq 5,$$

are SDs for $k = 1$ with treatments $N = 9$ and $2m$, respectively, and that

$$\mathbf{T} = \Omega(5, 0) + \Omega(5, 2) + \Omega(5, 5), \quad m = 5,$$
$$\mathbf{T} = \Omega(m, 2) + \Omega(m, m), \quad m \geq 6,$$

are SDs for $k = 2$ with treatments $N = 12$ and $m(m-1)/2+1$, respectively. For $\ell = 2$, Shirakura and Tazawa [29] showed that

$$\mathbf{T} = \Omega(6,0) + \Omega(6,3) + \Omega(6,5) + \Omega(6,6),$$
$$\mathbf{T} = \Omega(6,1) + \Omega(6,4) + \Omega(6,5) + \Omega(6,6),$$
$$\mathbf{T} = \Omega(6,0) + \Omega(6,1) + \Omega(6,4) + \Omega(6,5) \quad m = 6,$$
$$\mathbf{T} = \Omega(m,1) + \Omega(m,m-2) + \Omega(m,m-1) \quad m \geq 7,$$

are SDs for $k = 1$ with treatments $N = 28$ and $m(m+3)/2$, respectively, and that

$$\mathbf{T} = \Omega(6,1) + \Omega(6,2) + \Omega(6,4),$$
$$\mathbf{T} = \Omega(6,0) + \Omega(6,3) + \Omega(6,4) \quad m = 6,$$
$$\mathbf{T} = \Omega(7,1) + \Omega(7,4) + \Omega(7,7) \quad m = 7,$$
$$\mathbf{T} = \Omega(m,2) + \Omega(m,m-2) + \Omega(m,m) \quad m \geq 8,$$

are SDs for $k = 2$ with treatments $N = 36$, 43 and $m(m-1)+1$, respectively. Furthermore, they proved that the numbers of treatments N given in respective cases are minimum among SDs derived from BAs of strength $2(\ell+1)$ and m constraints. For the same parameter vectors ξ_1 and ξ_2 as in Ghosh [8], Shirakura and Tazawa [30] considered the situation for possible two nonzero interactions where there do not exist exactly two nonzero three-factor interactions. Furthermore, if there exists a nonzero three-factor interaction, then the other nonzero two-factor interactions may be of factors included in it, i.e., two nonzero interactions $\theta_{i_1 i_2}$ and $\theta_{j_1 j_2 j_3}$ are such that $\{i_1, i_2\} \subset \{j_1, j_2, j_3\}$. Shirakura and Tazawa [30] showed that for $m \geq 7$,

$$\mathbf{T} = \Omega(m,0) + \Omega(m,2) + \Omega(m,m),$$

is an SD for $k = 2$ with $N = (m^2 - m + 4)/2$ treatments.

For the same parameters vectors ξ_1 and ξ_2 as in Shirakura [23], let $\mathbf{T}_1$ be a plan of $N_1 = 2m+1$ treatments as $\mathbf{T}_1 = \Omega(m,0) + \Omega(m,1) + \Omega(m,m-1)$. By use of Hadamard matrices, Mukerjee and Chatterjee [18] characterized a plan $\mathbf{T}_2$ with N_2 treatments so that the design $\mathbf{T} = \mathbf{T}_1 + \mathbf{T}_2$ is an SD for $k = 2$ with treatments $N = 2m + 1 + N_2$. In fact, they constructed SDs for $k = 2$ with $N = 3m - 1$ if $m = 0 \pmod 4$, $N = 3m + 2$ if $m = 1 \pmod 4$, $N = 3m + 1$ if $m = 2 \pmod 4$, and $N = 3m$ if $m = 3 \pmod 4$. Shirakura, Suetsugu and Tsuji [25] considered more flexible choice for $\mathbf{T}_2$ instead of

Hadamard matrices in Mukerjee and Chatterjee [18]. For an $n \times m$ (0,1)-matrix $\mathbf{D}$ ($m \geq 4$), $\mathbf{D}$ is said to be an ST-array with n rows and m columns (ST-array (n, m)) if for every $n \times 4$ submatrix $\mathbf{D}_0$ of $\mathbf{D}$, there exist two rows in $\mathbf{D}_0$ such that the 2×4 matrix of the two rows has $(1,1)'$, $(1,0)'$, $(0,1)'$ and $(1,1)'$ as columns. For example, the following is an ST-array(6,8):

$$\begin{bmatrix} 0 & 1 & 0 & 0 & 1 & 0 & 1 & 1 \\ 0 & 1 & 1 & 0 & 0 & 1 & 0 & 1 \\ 0 & 1 & 1 & 1 & 0 & 0 & 1 & 0 \\ 0 & 0 & 1 & 1 & 1 & 0 & 0 & 1 \\ 0 & 1 & 0 & 1 & 1 & 1 & 0 & 0 \\ 0 & 0 & 1 & 0 & 1 & 1 & 1 & 0 \end{bmatrix}.$$

Shirakura, Suetsugu and Tsuji [25] showed that for $\mathbf{T}_1$ of Mukerjee and Chatterjee [18], $\mathbf{T} = \mathbf{T}_1 + \mathbf{T}_2$ is an SD for $k = 2$ with treatments $N = 2m + 1 + N_2$ if and only if $\mathbf{T}_2$ is an ST-array(m, N_2). Furthermore, they presented three methods for construction of an ST-array given below.

(i). Let $\mathbf{M}'(v \times b)$ be an incidence matrix of a balanced incomplete block design with parameters $v(\geq 4)$, b, r, k, and λ. If there exists at least one row of weight 2 for every $b \times 4$ submatrix of $\mathbf{M}(b \times v)$, then $\mathbf{M}^*$ obtained by deleting any row from $\mathbf{M}$ is an ST-array$(b-1, v)$.

(ii). Using properties of binary linear codes, an ST-array(n, m) of type of $n = r(r+1)/2$ and $m = 2^r$ can be constructed, where r is an integer greater than or equal to 3.

(iii). Using properties of quadratic residues over Galois field, an ST-array(n, m) of type of $n = m = 4t + 1 = q$, where t is an integer greater than or equal to 2 and q is a prime power.

In comparison between the above method (ii) and Mukerjee and Chatterjee's results, when $m = 8$, a plan $\mathbf{T}_2$ is with the same value $N_2 = 6$. However, when $m = 16$, 32 and 64, plans $\mathbf{T}_2$ of Shirakura, et al. [25] are with $N_2 = 10$, 15 and 21, whereas Mukerjee and Chatterjee's plans $\mathbf{T}_2$ are with $N_2 = 14$, 30 and 62, respectively. Also, for $m = 9$, a plan $\mathbf{T}_2$ with $N_2 = 9$ in the above third method is better for the number of N_2. Therefore, it is important that for a given $m(\geq 4)$, we obtain an ST-array(n, m) with a smaller value of n. For $m = 4, 5, 6, 7$ and 8, we have $n^* = 2, 4, 4, 6$ and 6, the minimum values of n for ST-arrays(n, m), respectively. As an open problem, it is interesting to note that the minimum value n^* is given for each $m \geq 9$.

For the noisy case $\sigma^2 > 0$, Srivastava [33] also proposed four procedures for solving the problem. One of them corresponds to the minimization of the sum of squares to error (SSE). Shirakura, Takahashi and Srivastava [26] studied stochastic properties of SSE for a given SD, and gave a probability that an SD could search the nonzero effects in ξ_2. Consider an SD $\mathbf{T}$ with N treatments. Suppose $\zeta(k \times 1)$ be a vector of nonzero effects of ξ_2. Model (1) reduces to

$$E(\mathbf{y}) = \mathbf{A}_1\xi_1 + \mathbf{A}_{21}(\zeta)\zeta, \tag{4}$$

where $\mathbf{A}_{21}(\zeta)$ is the $n \times k$ submatrix $\mathbf{A}_2$ of corresponding to ζ of ξ_2. Then the SEE $s(\zeta)^2$ can be written as

$$s(\zeta)^2 = \|\mathbf{y} - \hat{\mathbf{y}}\|^2 = \mathbf{y}'\big(\mathbf{I} - \mathbf{Q}(\zeta)\big)\mathbf{y},$$

where $\| \cdot \|$ is the norm of a vector and

$$\mathbf{Q}(\zeta) = \mathbf{A}(\zeta)\mathbf{M}(\zeta)^{-1}\mathbf{A}(\zeta),$$

where $\mathbf{A}(\zeta) = \big[\mathbf{A}_1 : \mathbf{A}_{21}(\zeta)\big]$ and $\mathbf{M}(\zeta) = \mathbf{A}(\zeta)'\mathbf{A}(\zeta)$.

Srivastava's procedure for searching nonzero effects is given below. For a given $\mathbf{y}$ in (4), calculate $s(\zeta)^2$ for each of the possible choices of ζ in ξ_2 (or $\mathbf{A}_{21}(\zeta)$ in $\mathbf{A}_2$). Let $\zeta_1(k \times 1)$ denote that subvector of ξ_2 for which $s(\zeta)^2$ turns out to be a minimum. Then, take ζ_1 as the possibly nonzero effect set of parameters. Of course, it is most desirable that the above ζ_1 is exactly equal to ζ_0 $(k \times 1)$, the vector of true nonzero effects of ξ_2. However, this is not ensured for the noisy case $\sigma^2 > 0$. So, for a vector of random variables $\mathbf{y}$, we may consider the probability

$$P = \min_{\zeta_0 \in \xi_2} \min_{\zeta \in \mathbf{A}(\xi_2;\zeta_0)} P\big(s(\zeta_0)^2 < s(\zeta)^2\big), \tag{5}$$

where $\mathbf{A}(\xi_2;\zeta_0)$ denotes the set of all possible $\zeta(\neq \zeta_0)$ of ξ_2. This means that the larger the value of P, the higher the confidence with which the true vector of parameters could be searched by the procedure. For the case of $k = 1$, let $\mathbf{a}(\zeta) = \mathbf{A}_{21}(\zeta)$, the $N \times 1$ column of $\mathbf{A}_2$ for ζ of ξ_2 in (4). Furthermore, let $r(\zeta) = \mathbf{a}(\zeta)'(\mathbf{I} - \mathbf{Q})\mathbf{a}(\zeta)$ and $\mathbf{b}(\zeta) = (\mathbf{I} - \mathbf{Q})\mathbf{a}(\zeta)/\big(r(\zeta)\big)^{1/2}$, where $\mathbf{Q} = \mathbf{A}_1'(\mathbf{A}_1'\mathbf{A}_1)^{-1}\mathbf{A}_1$. Suppose the components of the error vector $\mathbf{e}$ are distributed independently with a normal distribution $N(0, \sigma^2)$. Then, the result below was presented in Shirakura, Takahashi and Srivastava [26].

$$\begin{aligned} P\big(s(\zeta_0)^2 < s(\zeta)^2\big) &= 1 - \Phi\big(d(1-x)^{1/2}\big) - \Phi\big(d(1+x)^{1/2}\big) \\ &\quad + 2\Phi\big(d(1-x)^{1/2}\big)\,\Phi\big(d(1+x)^{1/2}\big), \end{aligned} \tag{6}$$

where $x = \mathbf{b}(\zeta)'\mathbf{b}(\zeta_0)$, $d = \left(r(\zeta_0)/2\right)^{1/2}\rho$ and $\Phi(x)$ is the distribution function of the standard normal distribution. Here, ρ is the actual quantity of unknown parameter ζ_0/σ. Shirakura, Takahashi and Srivastava [26] also gave a criterion for comparing SDs for $k = 1$ based on the search probabilities (5) and (6). Srivastava [34] proposed some criteria which are independent on the quantities of ζ_0 and σ^2. Ghosh [9] studied equivalence properties for the minimization of an SSE and also gave an estimator of the number k of nonzero effects in ξ_2.

In the class of designs of resolution V derived from BAs of strength 6, Shirakura and Ohnishi [24] presented optimal SDs ($6 \leq m \leq 8$) for $k = 1$ with respect to AD-criterion due to Srivastava [34], with N treatments satisfying ($m = 6$; $28 \leq N \leq 15$), ($m = 7$; $35 \leq N \leq 63$) and ($m = 8$; $44 \leq N \leq 74$). Anderson and Thomas [3], Chatterjee [4], and Chatterjee and Mukerjee [5, 6] treated and constructed SDs for general symmetric and asymmetric factorials.

5. Statistical design of scientific experiments

In the previous sections, we have reviewed the salient aspects of the theory of Search Designs (SD) and the Search Linear Model (SLM). We now discuss the subject at more fundamental levels, and offer some insights with respect to the information contained in an experiment. Also, we present a powerful method of using the SLM to analyze the data from any experiment, so as to determine the appropriate model and also a more accurate value of the location parameters under the same, relatively free from bias.

The purpose of the subject of Statistical Design of Scientific Experiments, briefly called Design of Experiments (DOE), is to collect data in such a way as to maximize the kind of information that is of interest to us. However, this requires that we are able to identify the model behind the phenomenon under study, and develop a measure of the information (contained in an experiment) on the parameters in the model. Hence, at the fundamental levels, the problems faced in DOE are to identify the model as precisely as possible, and estimate parameters under the model as precisely as possible. Also, in doing all this, we shall have to cope with measurement errors and errors due to random fluctuations, caused by the fact that a large number of factors (each with a *small* effect on any observation), which are unknown and beyond our control, are also influencing the observations that we make on the phenomenon under study.

Consider a phenomenon. Typically, one would have a set of guesses about what would be an appropriate model to describe the phenomenon. Here we shall restrict ourselves to situations where a linear statistical model would be adequate. [However, non-linear models are also quite common (particularly, in the more exact sciences like physics and chemistry), and it would be quite worthwhile to extend the methods of this paper to the non-linear case.] So, let us assume that we have q models to begin with. Let $\mathbf{y}(n \times 1)$, with elements $y_1, y_2, \ldots, y_n$, be a vector of some set of n observations on the phenomenon. Let the q models respectively be

$$(7) \qquad E(\mathbf{y}) = \mathbf{X}_i\beta_i, \qquad i = 1, 2, \ldots, q;$$

where E denotes expected value, the $\mathbf{X}_i(n \times p_i)$ are known matrices of real numbers, and where the ith model involves p_i unknown parameters which are the elements of the vector $\beta_i(p_i \times 1)$. The matrix $\mathbf{X}_i$, often called the *design matrix,* comes from the postulates that we make concerning the structure of the phenomenon and how the observations y_j $(j = 1, 2, \ldots, n)$ fit into this structure. For simplicity of discussion, throughout this paper, we shall assume that the n observations coming from any experiment under consideration are all independent.

It is important to remark that, in the physical sense, the experiment (say, E_x) has three components: (a) The selection of the n units (say, U), (b) The action taken on each individual unit, and (c) The observation obtained from U. A very large part of DOE deals with *comparative experiments,* where we have a set of (say, v) treatments (these being, for example, v varieties of wheat, which are being compared with respect to their yield). Part (b) above, namely the *action taken,* corresponds to the application of one of the treatments to a unit. In *factorial experiments,* the set of treatments may be a set of level-combinations of a set of factors. For example, we may have 4 factors (say, Nitrogen (F1), Phosphorus (F2), Potassium (F3), and Manure (F4)), each at three levels 0 (None), 1 (20 pounds per acre), and 2 (40 pounds per acre). Then, we get $3^4 = 81$ level-combinations, which may be denoted by vectors (t_1, t_2, t_3, t_4), where t_j $(j = 1, 2, 3, 4)$ denotes the level of the jth factor, and takes the values 0, 1, and 2. In an actual experiment with these four factors, all treatments may not be used, and v may be a much smaller number.

The experiment E_x (on the phenomenon under consideration) does not give the model or even the guessed set of models. Such models are only a kind of superimposition of our thought process on E_x. Thus, the models under consideration above could have been partially or wholly thought of before, during, or after the experiment. Also, the author believes that the process of the selection of U does not come into picture, so long as it is unrelated to the action taken on the units. Thus, suppose that P_1 and P_2 are two processes of selection of units, such that U could get selected from any one of them (even though the probability that U is selected under P_1 may be quite different from that under P_2). Then, according to the author, P_1 and P_2 etc. do not form a component of the experiment; only U (the end result of the selection process used) does. (This point shall be elaborated further later on.)

The main problem of interest here is how to identify the model that *seems to be* the closest to the phenomenon. Notice that we use the phrase *seems to be* because we can never be absolutely certain of everything concerning a phenomenon. Our attempt is to find a model that describes the phenomenon reasonably well from the practical viewpoint.

6. The information provided by an experiment

It will be interesting here to mention some conversations that the author had (in the late 1960s) with the great mathematician Alfréd Rényi, after whom the Mathematical Institute of the Hungarian Academy of Sciences is now named. Rényi was also a probabilist and information theorist; his book on Probability Theory (with a chapter at the end on Information Theory) is a classic. He defined a measure of information, later called the *Rényi information,* which is a generalization of the Shannon information measure. Later, Srivastava [32] discovered that the Rényi information can be amazingly used to estimate the dimensionality d_2 of a (relatively large) set of data points which are defined in d_1 dimensions, but which actually lie on a (possibly curved) manifold of a lower dimension d_2. For example, we may have $d_2 = 2$, $d_1 = 6$, and the data points could be vectors $\mathbf{z}$ with elements $z_1, \ldots, z_6$, where $z_1 = x_1 + x_2$, $z_2 = {x_1}^3 + {x_2}^2$, $z_3 = \log(1 + {x_1}^4 + {x_2}^2)$, $z_4 = \exp(-x_1 - x_2)$, $z_5 = \tan(z_1 - z_2)$, and $z_6 = (x_1 + 3x_2)/(1 + 5x_1 + x_2)$, where x_1 and x_2 vary (say, over a 100 equidistant points in the interval $(0, 99)$. With 10,000 points so generated, the estimate of d_2 should be close. But, even a subset of say, 3,000 randomly selected points may also work. Note that, here, the researcher has only the points z, and has no knowledge at all as to how many x's are involved (if they are involved), and in what way the involvement occurs, i.e., what functions are involved. The technique based on the Rényi information tends to *see through* whatever functions may be involved. However, as interested readers would find, more work in that field needs to be done. The Rényi information uses additional parameters, and J. N. Srivastava believes that these parameters themselves carry various kinds of information. But, this needs to be explored more.

Both Rényi and Srivastava occasionally visited with R. C. Bose at the University of North Carolina, Chapel Hill, in those years. Often, they used to have lunch together that would continue into discussions of fundamental questions. Once, Rényi asked Srivastava what he was currently working on. *Optimal design theory,* was the reply. Rényi was not well acquainted with DOE, so the author gave him an introduction. Srivastava explained that in optimal design theory, we assume that the model is known, and we try to select a set of treatments (from the total set of possible treatments) such that we obtain the maximum amount of *information* on the location parameters (which would correspond to the vectors β_i in the ith model in (7)). Rényi pointed out that if we take the normal distribution for example, the location parameters do not enter into the expression associated with its entropy. From that angle, all designs are equally informative. Srivastava mentioned that in DOE, we use the *Fisher Information* (which would correspond to the matrix $\mathbf{X}_i'\mathbf{X}_i$ in the ith model in (7)). However, still the fact remained.

Another time, Rényi explained that he was working on *Search theory,* which could also be probabilistic. A very simple example of the same would be the following problem. There are $(n_1 + n_2 + n_3)$ coins all of which look alike, but which are divisible into three groups (of n_1, n_2, and n_3 coins respectively) with respect to weight. The weight of each coin in any group is the same, but coins belonging to different groups have different weights. The values of the n's are not known. We have an accurate balance in which we can put coins only to compare. What is the minimum number of weighings needed to separate out the coins correctly into their groups?

Rényi's interesting presentation of search problems led Srivastava to realize that in factorial design theory as well, there are hidden search problems. Thus, consider a 2^m factorial experiment (in which there are m factors each at 2 levels) with level combinations $(t_1, t_2, \ldots, t_m)$ where the t_i equal 0 or 1. Let $a(t_1, t_2, \ldots, t_m)$ denote the *true* yield corresponding to the treatment $(t_1, t_2, \ldots, t_m)$, where the true yield is what one would observe if there are absolutely no other causes that influence the yield of a unit except the treatment applied to it. Then, as explained earlier in this paper, the parameters of interest are also 2^m in number, these being the general mean μ, the main effects A_i, the 2-factor interactions A_{ij}, and the higher order interactions. It is an empirical fact that almost all the time, a lot of the higher order effects are negligible. However, almost always, it is also a fact that at least a few of the interactions are not known in the sense that not only we do not know the value of these interactions, we do not even know their identity, i.e., we do not know which interactions are the non-negligible ones. In view of this fact, Srivastava realized that in factorial experiments there is a hidden search problem, namely, we have to search (i.e., identify) the non-negligible interactions. Thus, the Theory of Search Linear Model and Search Designs were born. The first paper (Srivastava [33]) was presented in a conference in March 1973.

It should be emphasized that the SLM is based on the real nature of the scientific fields. Many workers in the discipline (or, the lack thereof) of DOE still pretend that only the main effects (and, sometimes, perhaps a few interactions known to the experimenter) need to be estimated. The so-called main-effect plans are still popular. Some people wrongly call them *screening designs,* because they believe that by using them they can screen out the *small* factors (i.e., those with small main effects), so that later they can concentrate on the *big* ones. But, their *screening design* would work only if it so happens that the interactions that are non-negligible are not confounded with the main effects of the small factors. Since we do not know which factors are small and which interactions are non-negligible, the *screening* could give correct answer only as a fluke. Moreover, one would not know if the *screening* gave a right or a wrong answer.

Also, the basic assumption that if a factor is small, then its interactions will be small too is not necessarily true. In some fields, it is always false. An example is the field of nutrition experiments, with enzymes as factors. Since enzymes act as a catalyst alone, they have large (sometimes, quite high-ordered) interactions, but their *main-effects* are small.

The assumption that interactions are negligible and a so-called *main effect plan* is sufficient, is also risky and misguided. This can be verified, as Srivastava did in 1976 by looking into journals (in the field of agriculture, and also social sciences) containing papers describing the results of full experiments. Srivastava found interactions present almost always, sometimes even involving several factors. In certain sectors of industry, full experiments have not been done. Dependence on *main effect plans* has resulted in a confounded situation, where one believes one has reached Quality, but where a lot of further improvement is still possible. Society can be helped by the creation of consulting companies who understand the foundations of DOE, and who can educate the users and lead them out of obsession with elementary DOE (like the Plackett-Burman plans and rotatable designs). It may be added that some of the tools used are simply wrong; this includes, for example, Addelman's (1962, 1963) so-called *orthogonal main effect plans* which had been wrongly included even in some books. (See Srivastava and Ghosh [44].)

The new companies and researchers can take up the ideas presented here and create a revolution in DOE; this is a lucrative proposition since it shall benefit all concerned, simply because it shall create material wealth.

Coming back to the conversation with Rényi, Srivastava feels that his remark that each design (with the same number of observations) would have the same amount of information, is fundamentally justified. Indeed, Srivastava feels that such competing designs really differ only in the kinds of information that they give. Indeed, it seems that we have

$$I = I_1 + I_2 + I_3 + I_4, \tag{8}$$

where I denotes the total information in the observations from the experiment, I_1 denotes the information that the experiment provides concerning competing models for explaining the observations, I_2 denotes the information that the experiment provides on the parameters within a model (assuming that the model is true), I_3 provides information on the random fluctuations or measurement errors associated with the observations on the experimental material used, and I_4 is some *miscellaneous* information whose nature is yet to be understood. So, it seems that while I is the same, the $I_1, \ldots, I_4$ may vary from design to design. Srivastava believes that I could be a simple function of n itself. Perhaps the information measure could be standardized so that we have $I = n$.

Optimal design theory assumes that the model is known, and tries (in the sense of Fisher information) to maximize the part of I_2 that deals with location parameters. Also, I_3 is linked to what has been called *revealing power,* which is a measure of the ability of a design to identify the correct model out of a set of competing models. Similarly I_1 is linked to what has been called *Sensitivity.* If the *error variance* (which is a measure of random fluctuations acting on the units, and of measurement error) is large, the design is less sensitive, and vice versa. Finally, I_4 may be linked to the information contained in the design about the set of competing models under consideration relative to the situation of *no models* (i.e., total ignorance).

Which of these four kinds of information we need more in a given situation? The answer to this important question depends upon what stage we are in with respect to our enquiry concerning the phenomenon, i.e., what kind of, and how much, information we already possess on the phenomenon. If we believe there is too much random fluctuation and measurement error, it may be necessary to increase n and replicate the observations more. If we believe we know the model adequately except for the parameters, then optimal design theory will be important. Otherwise, if the model is not known, but we have good guesses, we need a design with high revealing power. If even good guesses are not available, we need to try to theoretically understand the phenomenon more before venturing out into experimentation.

As yet, no class of information measures has been proposed which would validate the equation (8) in a cogent manner, although this equation does appear to hold the truth. One basic characteristic of DOE is that although the observations usually have continuous distributions, the set of observations is (almost always) discrete. Thus, there is a combination of continuous and discrete problems that must be coped with. However, there is a lacuna between the information theory dealing with the continuous and the discrete situations. This lacuna must be bridged. Also, we must combine the concept of *information* in an observation coming from a distribution with the concept of *information on a particular set of parameters coming from the distribution.* Finally, how much information is there when there is total ignorance of the model?

DOE is a part of the larger field of Information Sciences, which has Information Theory as its nucleus. It holds the promise of being evergreen, since the author believes that it probably subsumes a large chunk of Physics as well. The problems mentioned in connection with (8) above are wide open, challenging, and inviting.

7. Some elegant combinatorial problems arising in search designs

In the first paper, the SLM defined in (1) was considered. For simplicity of discussion, we assume that the elements of $\mathbf{y}$ are all independent with the same variance σ^2. The matrix $\mathbf{A}_1$ is $(n \times \nu_1)$ and $\mathbf{A}_2$ is $(n \times \nu_2)$; both of these matrices are known. Also, ξ_1 is a vector of ν_1 parameters which are all unknown, and ξ_2 is a vector of ν_2 parameters which are all unknown except that it is known that at most k of them are nonzero. Notice that this model is a special case of (7), with

$$q = \sum_{j=0}^{k} \binom{\nu_2}{k-j}, \tag{9}$$

where the different models are obtained by merging ξ_1 with the set of parameters of ξ_2 that are assumed to be possibly nonzero. The model (1) has been discussed extensively in earlier sections of this paper, particularly for the case when the y's are observations on the level-combinations from the 2^m factorial experiment, and the parameters on the right side of (1) are the various main effects, and interactions etc. in some order.

There is a further special case of the last mentioned situation where $\nu_1 = 0$, and $\nu_2 = 2^m$, in which case, for simplicity, we shall denote $\mathbf{A}_2$ simply by $\mathbf{A}$. In that case, certain sub-problems of the general problem lead to the following problem. Let $EG(m, 2)$ denote the finite Euclidean space of m dimensions based on $GF(2)$, the finite field with two elements. Let $\mathbf{T}$ be a set of points in $EG(m, 2)$. Then, $\mathbf{T}$ is said to be a t-covering of $EG(m, 2)$ if $\mathbf{T}$ has a non-empty intersection with every $(m-t)$-flat of $EG(m, 2)$. Given m, we need to find t-coverings $\mathbf{T}$ that are minimal in size, i.e., we need to find a $\mathbf{T}$ such that the number of points in $\mathbf{T}$ is a minimum.

The paper by Katona and Srivastava [17] presents some basic results on the case $t = 2$. Suppose $\mathbf{T}$ has n points. Then, this paper also looks at the problem from the reverse angle. We ask the question: Given n, what is the largest value of m such that there exists a $\mathbf{T}$ with n points that is a t-covering of $EG(m, 2)$. Given n, bounds are presented on m, and connections with coding theory are made.

Now, we can consider $\mathbf{T}$ presented as a $(0, 1)$-matrix of size $(n \times m)$, such that the rows represent some level-combinations from the 2^m experiment, and the columns represent the m factors. Let $\mathbf{T}^*$ be the $(n \times m)$ matrix obtained from $\mathbf{T}$ by interchanging zeros and ones. Let $\mathbf{Z}$ be the $(n \times 2^m)$ matrix obtained from $\mathbf{T}$ by taking as columns all the vectors (not necessarily distinct) generated by taking all possible linear combinations of the columns of $\mathbf{T}^*$.

Next, suppose that the elements of $\mathbf{y}$ correspond, in order, to the level-combinations in the rows of $\mathbf{T}^*$. Then, the matrix $\mathbf{A}$, mentioned earlier in this section, is obtainable (except possibly for the order of the columns) from $\mathbf{Z}$ by replacing everywhere 0 by 1, and 1 by (-1). Notice that while $\mathbf{Z}$ is over $GF(2)$, the matrix $\mathbf{A}$ is over the real field.

A matrix $\mathbf{G}$ (over any field F) is said to have the property P_t with respect to columns if and only if every set of t columns of $\mathbf{G}$ is linearly independent (over F). The SLM leads us into the situation where we are interested in the matrix $\mathbf{A}$ to have the property P_t over the real field for the smaller values of t. The case $t = 4$ corresponds to 2-coverings of $EG(m, 2)$, and has been studied in more detail. Also, there is unpublished work on $t = 6$, largely by Katona.

Consider (1) again. The result in (2), discussed in detail in the other parts was established in Srivastava [33]. A Search Design (SD) is a design such that if the corresponding model is written in the form (1), then the $\mathbf{A}$ matrices occurring there would satisfy (2). As is seen in the previous parts, a lot of effort has been made to build SD's, which is certainly an important goal. However, in actual applications, it turns out that the condition (2) is relatively too stringent if it is looked at only superficially. We elaborate this in the next paragraph. For simplicity, we shall assume that σ^2 is zero. If this condition is not met (which would be expected in practice in statistical situations), then our assertions will become probabilistic instead of being deterministic.

Suppose (2) does hold. Then, it guarantees that whatever $\mathbf{y}$ the model (1) may produce, it would be such that for that particular $\mathbf{y}$, there will be a unique solution to (1) with a unique set of parameters, each of which shall have a unique value. On the other hand, if (2) does not hold, then the model (1) may throw up a $\mathbf{y}$ such that there may be more than one (possibly, an infinite number of) solutions for the set of parameters on the right side of (1). However, it is NOT saying that if (2) does not hold, then every value of $\mathbf{y}$ shall land us into a situation where more than one parameter sets shall be found to be mixed up. Indeed, only those values of $\mathbf{y}$ shall be in this mess, which are simultaneously linear combinations of two (or more) distinct sets of columns of $[\mathbf{A}_1, \mathbf{A}_2]$, where each set has at most k columns of $\mathbf{A}_2$. But, these linear combinations will have to have a given fixed set of coefficients. To elaborate further, suppose that we have

$$(10) \qquad \mathbf{A}_{21}\theta_1 = \mathbf{A}_{22}\theta_2 = \mathbf{z}_1, \quad \mathbf{A}_{21}\theta_2 = \mathbf{z}_2, \quad \nu_1 = 0,$$

where $\mathbf{A}_{21}$ and $\mathbf{A}_{22}$ are distinct sub-matrices of $\mathbf{A}_2$ with k columns each. Then, in the noiseless case if $\mathbf{y} = \mathbf{z}_1$, a unique solution to (1) will not be there. However, if $\mathbf{y} = \mathbf{z}_2$, and there is no sub-matrix of $\mathbf{A}_2$ other than $\mathbf{A}_{21}$

on whose columns $\mathbf{z}_2$ is dependent, then there will be a unique solution to (1) even though (2) does not hold.

The moral of the discussion in the last paragraph is that very often, there can be a $\mathbf{y}$ for which (1) can have a unique solution even though (2) does not hold. In other words, if the equation (2) holds, then a unique solution is guaranteed for all $\mathbf{y}$ that we are likely to observe. But, if (2) does not hold then, we may possibly obtain a $\mathbf{y}$ for which there is no unique solution. However, it would be clear to a reader acquainted with the subject, that in the last situation, we should usually be able to resolve the problem by trying one, or may be a few, more treatments. This idea is what gives rise to multistage design procedures, where we run a set of treatments first, which is possibly followed by another set, which may further be followed by a third set, and so on. (See Srivastava and Chu [41]).

The discussion in the last two paragraphs encourages us to consider designs that do not satisfy (2), but which would still have a high revealing power, particularly when we use the methods of analysis mentioned in the next section.

As the material in the other parts of this paper shows, in the construction of SD's, a lot of investigations have been made for the case where there is a relatively large amount of symmetry with respect to factors. In other words, the *balanced* case has been studied quite often. However, the *partially balanced* approach seems to hold much more promise. This is exemplified by the 2^8 design in 24 assemblies described in Srivastava [36]. Here, the vertices of a 3-dimensional cube are put in correspondence with the 8 factors, and 12 out of the 24 runs correspond to the edges. Three runs are chosen by *inspection,* and do not conform to any kind of *balance.* Relatively speaking, this design is able to identify more interactions than most others in the literature.

8. Analysis under the general SLM using simulation distributions

Consider the general SLM (7). It should be emphasized that in all the m models represented in (1), the set of the n experimental units and the observations $y_1, \ldots, y_n$ thereon are all fixed entities. As mentioned before, the models only represent alternate parametric structures imposed on these entities, and such imposition can therefore be done before, during, or after the experiment. The problem, of course, is to determine the most suitable model. In this section, we propose a general method for doing the same. Special cases of this procedure, suggested earlier by the author, have been used with impressive results.

Since the matrices $\mathbf{X}_i$ are known for all i, let us consider the ith model $(i = 1, \ldots, q)$ individually. We can compute the sum of squares due to error under the ith model; dividing this by the degrees of freedom due to error, we obtain the mean square due to error (denoted herein by ${\psi_i}^2$). Without loss of generality, we shall say that the model for $i = 1$ is *special;* how to choose the special model will be discussed later on. Let

$$\min \psi^2 = \min_{i=2\ldots q} {\psi_i}^2 \tag{11}$$

$$R = \left[{\psi_1}^2 / \min \psi^2\right]. \tag{12}$$

For ease of explanation and understanding, we proceed with examples given below.

Example 1. Consider a 4×4 Latin Square (LS). The observations come from $n (= 16)$ units. The LS corresponds to the *special model,* there being 1 degree of freedom (d.f.) for the general mean, 3 each for rows, columns, and treatments, and 6 for error. Also, suppose there are t non-additive (NA) cells, which means that these cells have their own (possibly large) contribution to the observation made on them, which therefore does not obey the usual linear model associated with a LS. If a cell is NA, the observation on it is clearly useless, and may thus be ignored. For simplicity, let us first take $t = 1$. Since any of the 16 cells can be NA, 16 models arise here, each model having 15 observations leaving one of the 16 cells aside. Then, we shall have $q - 1 = 16$, or $q = 17$. The special model would correspond to the customary analysis of the LS design. Now, suppose an experiment is actually done and we do have the y values. Firstly, suppose there are no NA cells. Then, we shall expect ${\psi_i}^2$ to be of the same order for each i. However, if there is a NA cell, then the model obtained by excluding the same should be expected to give a smaller error mean square than the model for $q = 1$. The reason is that the special model tries to *absorb* the value of the NA cell under its system, by treating it as a random fluctuation.

Hence, if there is a single NA cell actually present, we shall expect the value of R to be much larger than what it would be if there was no NA cell at all. How can we determine if the observed value of R is too large? We need to know the distribution of R under the assumption that there are no NA cells at all. Theoretical determination of this distribution would at best be messy. But, in the modern computer age, we can easily obtain it by simulation. We generate, say, 5000 samples, each sample throwing up a value of y by choosing some value of the general mean and of the row, column, and treatment effects, and also by putting in some value for the random fluctuation (on each unit) by generating samples from the assumed distribution of the y's (apart from the location parameters). In the experience of the author, a sample of 5000 should give quite a smooth distribution, against which the observed value of R can be checked.

In Srivastava and Wang [46], such a procedure was applied to real data from several books, and a very large amount of non-additivity was found in four out of nine cases. In some cases, the value of R was so large that it was quite outside the range observed in the simulation distribution. To evaluate the probability of observing a value of R that large, recourse was made to the Tchebycheff inequality.

In the above, we talked about one NA cell in a 4×4 square. However, in one of the four non-additive situations mentioned in the last paragraph (from the book Bliss (1967)), we found 3 NA cells. How did we find three NA cells? Well, we obtain R using one NA cell as described above. Then, we considered $\binom{16}{2}$ models in each of which, 2 cells out of the 16 were ignored. We found an appreciably decreased value of R. We next tried $\binom{16}{3}$ more models in which 3 cells were ignored, and the value of R reduced further quite a bit. But, when we tried ignoring four cells, R was about the same as for the case of 3 cells. The probability of the observed value of R (using 3 cells) was about 0.000000373.

It should be emphasized here that it would not be wise to simply discard the NA cells, once they are located. A wise experimenter should study as fully as possible why that non-additivity arose, and what does it signify. May be it is telling us some thing more important than the rest of the experiment!

Example 2. It is clear that the above methods can be easily generalized to various kinds of situations. Thus, for example, the row column designs (of which the LS is a special case) have been discussed in Srivastava and Wang [46]. An extensive study of LS showed that the probability that NA cells that are only moderate large (i.e., those which have moderately large values of the non-additivity parameter) can be correctly identified is disappointingly small, whereas the occurrence of non-additivity appears to be abundant. The moral is that it is risky to use large (say, more than 4 rows or columns) designs. The alternative is to use NMBDs (Nested Multidimensional Block Designs, introduced in Srivastava and Beaver [40]) that are a generalization of Lattice Squares.

For example, if we have 5 treatments, then instead of using a 5×5 LS which has 5 replications, we can use the smaller design with 4 replications that has 5 2-dimensional blocks, each block being of size 2×2 in which 4 treatments are tried in a row column design. Similarly, for 9 treatments there is the nice *balanced Lattice Square* design in four replications, in which there are 4 2-dimensional blocks, each block having all the 9 treatments arranged in a 3×3 row-column design, the arrangements corresponding to the planes of $EG(2, 3)$.

It would be interesting to recall an anecdote in this connection. When the author first met with Professor R. C. Bose (18 October 1959), Bose urged him to try to make at least three pair-wise orthogonal 10×10 LSs. However, the author rejected the idea, partly on statistical ground, because Fisher and Yates had said that there may be non-additivity in large row column designs, and size 10 was certainly quite large. This meant that 10×10 LSs were not appropriate. Bose accepted the reasoning. However, it still remained to show that the hunch of Fisher and Yates was correct. The author wished to investigate, particularly after the idea of SLM was developed. However, only in 1990, he could bring full attention to the same. The results of Srivastava and Wang support Fisher and Yates, but that non-additivity would turn out to be the hazard that it is was not anticipated until 1995. It is to be emphasized that we are not saying that row-column designs be abandoned; rather they should be used in smaller sizes, as is possible using NMBDs.

Example 3. Consider the 2^m factorial experiment along with (1). In ξ_1, put all parameters that are such that there is a relatively strong possibility that they will be non-negligible. In ξ_2, put all the remaining parameters about which we are not (relatively) certain that they are negligible. Let the special model be the one that takes $k = 0$. With these specifications, proceed as explained in the beginning of the section, taking k respectively equal to $1, 2, 3, \ldots$, and so on, and observing R (for each value of k), and stopping when the next larger value of k does not decrease R appreciably.

If we believe that the number of non-negligible elements in ξ_2 is at least k^*, and the design does not satisfy the condition (2) (with $k = k^*$), we should still go ahead and use the above procedure. Because of the remarks made in the paragraphs following (5), there is a good chance that the non-negligible parameters would be successfully identified, provided that under the special model, there are a lot of degrees of freedom which are for error but not for *pure error.* (Sometimes, this can be achieved by a multistage design.) This point cannot be delineated further both because of lack of space and because further research is probably needed on many crucial questions. But, for prospective researchers, it is a good line to investigate, since the method is very general and very powerful.

Example 4. We now consider another kind of example which would be useful at the preliminary stages of arranging experimental material for future experimentation. We discuss what is called a uniformity trial. For simplicity, an agricultural setting will be assumed. Suppose there is a $u \times v$ rectangle of uv plots, on which a uniformity trial is conducted. The problem is to ascertain if row effects are present, and / or column effects are present, and furthermore, what is the incidence of the NA cells.

For this problem, the special model does not include any parameters, except one general mean, which is the expected value of all observations. Other models could include one with row effects only, one with column effects only, one with both row and column effects. Including the special model, this totals to four models so far. From each of these four models, we can generate uv new models by taking one cell as NA. We can generate $\binom{uv}{2}$ further models by taking 2 cells as NA, and so on.

The method of this section can then be applied, and the situation can also be explored.

Remark 1. It should be noted that it is possible that in a given situation, the value of R (relative to the simulation distribution) may not be found to be too large, and yet there may some non-negligible parameters still hidden in the system. This can happen if the system does not have *degrees of freedom* to allow for a more extensive search. The unidentified non-negligible parameters could be large, and may boost both the numerator and the denominator of R, causing the value of R to be moderate. Further research is needed to clarify how to proceed so that enough *degrees of freedom* are available. The multistage design procedure would be quite promising in this regard.

REFERENCES

[1] H. Akaike, Fitting autoregressive models for prediction, *Ann. Inst. Staist. Math.*, **21** (1969), 243–247.

[2] H. Akaike, Information theory and an extension of the maximum likelihood principle, in: *2nd International Symposium on Information Theory* (ed: B. N. Petrov and F. Csaki), Akadémiai Kiadó (Budapest, 1973), 267–281.

[3] D. A. Anderson and A. M. Thomas, Weakly resolvable IV.3 search designs for the p^n factorial experiments, *J. Statist. plann. Inference,* **4** (1980), 299–312.

[4] K. Chatterjee, Search designs for searching for one among the two- and three-factor interaction effects in the general symmetric and asymmetric factroials, *Ann. Inst. Statist. Math.,* **42** (1990), 783–803.

[5] K. Chatterjee and R. Mukerjee, Some search designs for symmetric and asymmetric factorials, *J. Statist. plann., Inference,* **13** (1986), 357–363.

[6] K. Chatterjee and R. Mukerjee, Search designs for estimating main effects and searching several two-factor interactions in general factorials, *J. Statist. Plann. Inference,* **37** (1993), 385–392.

[7] S. Ghosh, On main effect plus one plans for 2^m factorials, *Ann. Statist.,* **8** (1980), 922–930.

[8] S. Ghosh, On some new search designs for 2^m factorial experiment, *J. Statist. plann. Inference,* **5** (1981), 381–389.

[9] S. Ghosh, Influential nonnegligible parameters under the search linear model, *Commun. Statist. Theory Methods.,* **16** (1987), 1013–1025.

[10] S. Ghosh, Sequential assembly of fractions in factorial experiments, in: *Design and Analysis of Experiments: Handbook of Statistics* **13** (ed: S. Ghosh and C. R. Rao), Elsevier (Amsterdam, 1996), 407–435.

[11] S. Ghosh and C. Burns, Comparison of four new general classes of search designs, *Australian & New Zealand J. of Statist.,* **44(3)** (2002), 357–366.

[12] S. Ghosh and C. R. Rao, *Design and Analysis of Experiments,* North-Holland, Elsevier Science B.V. (Amsterdam, 1996).

[13] S. Ghosh and H. Talebi, Main effect plans with additional search property for 2^m factorial experiments, *J. Statist. Plann. Inference,* **36** (1993), 367–384.

[14] S. Ghosh and T. Teschmacher, Comparisons of search designs using search probabilities, *J. Statist. Plann. Inference,* **104** (2002), 439–458.

[15] B. C. Gupta, A bound connected with 2^6 factorial search designs of resolution 3.1., *Commu. Statist.-Theor. Meth.,* **17(9)** (1988), 3137–3144.

[16] B. C. Gupta and S. S. R. Carvajal, A necessary condition for the existence of main effect plus one plans for 2^m factorials, *Commu. Statist.-Theor. Meth.,* **13(5)** (1984), 567–580.

[17] G. Katona and J. N. Srivastava, Minimal 2-coverings of a finite affine space based on $GF(2)$, *J. Statist. Plann. Inference,* **8** (1983), 375–388.

[18] R. Mukerjee and K. Chatterjee, An application of Hadamard matrices for the construction of main effect plus two plans for 2^m factorials, *Utilitas Math.,* **45** (1994), 213–218.

[19] T. Ohnishi and T. Shirakura, Search designs for 2^m factorial experiments, *J. Statist. Plann. Inference,* **11** (1985), 241–245.

[20] R. L. Plackett and J. P. Burman, The design of optimum multifactorial experiments, *Biometrika,* **33** (1946), 305–325.

[21] T. Sawa, Information criteria for discriminating among alternative regression models, *Econometrica,* **46** (1978), 1273–1282.

[22] T. Shirakura, Main effect plus one or two plans for 2^m factorials, *J. Statist. Plann. Inference,* **27** (1991), 65–74.

[23] T. Shirakura, Fractional factorial designs of two and three levels, *Discrete Math.,* **116** (1993), 99-135.

[24] T. Shirakura and T. Ohnishi, Search designs for 2^m factorials derived from balanced arrays of strength $2(\ell + 1)$ and AD-optimal search designs, *J. Statist. Plann. Inference.,* **11** (1985), 247–258.

[25] T. Shirakura, T. Suetsugu, T. and Tsuji, Construction of main effect plus two plans for 2^m factorials, *J. Statist. Plann. Inference,* **105** (2002), 405–415.

[26] T. Shirakura, T. Takahashi and J. N. Srivastava, Search probabilities for nonzero effects in search designs for the noisy case, *Ann. Statist.,* **24** (1996), 2560–2568.

[27] T. Shirakura and S. Tazawa, Enumeration a representation of nonisomorphic weighted graphs, *Kobe J. Math.,* **3** (1986), 229–235.

[28] T. Shirakura and S. Tazawa, Series of main effect plus one or two plans for 2^m factorilas when three-factor an higher order interactions are negligible, *J. Japan Statist. Soc.,* **21** (1991), 211–219.

[29] T. Shirakura and S. Tazawa, A series of search designs for 2^m factorial designs of resolution V which permit search of one or two unknown extra three-factor interations, *Ann. Inst. Statist. Math.,* **44** (1992), 185–196.

[30] T. Shirakura and S. Tazawa, Search designs for balanced 2^m factorial designs of resolution $2\ell+1$ when $(\ell+s+1)$-factor and higher order interaction are negligible, *J. Comb., Infor., Sys. Sci.,* **18** (1993), 233–246.

[31] J. N. Srivastava, Some general existence conditions for balanced arrays of strength t and 2 symbols, *J. Comb. Theory,* **13** (1972), 198–206.

[32] J. N. Srivastava, Application of the information function to dimensionality analysis and curved manifold clustering in: *Proceedings of the third International Symposium on Multivariate Analysis* (ed. P. R. Krishnaiah), Academic Press (New York, 1973), 369–381.

[33] J. N. Srivastava, Designs for searching non-negligible effects, in: *A Survey of Statistical Design and Linear Models* (ed. J. N. Srivastava), North-Holland, Elsevier Science B.V. (Amsterdam, 1975), 507-519.

[34] J. N. Srivastava, Optimal search design, or designs optimal under bias free optimality criteria, in: *Statistical Decision Theory and Related Topics II* (eds. S. S. Gupta and D. S. Moore), Academic Press (New York, 1977), 375–409.

[35] J. N. Srivastava, Sensitivity and revealing power: Two fundamental statistical criteria other than optimality arising in discrete experimentation, in: *Experimental Designs, Statiscical Models, and Genetic Statistics Models, and Genetic Statistics* (ed. K. Hinkelmann), Marcel Dekker (New York, 1984), 95–117.

[36] J. N. Srivastava, A 2^8 factorial main effect plus plan with good revealing power, *Sankhya,* **54** (1992), 461–474.

[37] J. N. Srivastava, Nonadditivity in row-column designs, *J. Combin, Inf. Sys.,* **18**, 85–96.

[38] J. N. Srivastava, A critique of some aspects of experimental design, in: *Design and Analysis of Experiments: Handbook of Statistics* **13** (eds. S. Ghosh and C. R. Rao), Elsevier (Amsterdam, 1996), 309–341.

[39] J. N. Srivastava and S. Arora, An infinite series of resolution III.2 designs for the 2^m factorial experiment, *Discrete Math.,* **98** (1991), 35–56.

[40] J. N. Srivastava and R. J. Beaver, On the superiority of the nested multidimensional block designs, relative to the classical incomplete block designs, *J. Statist. Plann. Inference,* **13** (1986), 133–150.

[41] J. N. Srivastava and J. Y. Chu, Multistage design procedures for identifying two factor interactions, when higher effects are negligible, *J. Statist. Plann. Inference,* **78** (1999), 149–189.

[42] J. N. Srivastava and S. Ghosh, A series of balanced 2^m factorial designs of resolution V which allow search and estimation of one extra unknown effect, *Sankhya (Ser. B),* **38** (1976), 280–289.

[43] J. N. Srivastava and S. Ghosh, Balanced 2^m factorial designs of resolution V which allow search and estimation of one extra unknown effect, $4 \leq m \leq 8$, *Commu. Statist.-Theor. Meth.,* **6** (1977), 141–166.

[44] J. N. Srivastava and S. Ghosh, On non-orthogonality and non-optimality of Addelman's main effect plans satisfying the condition of proportional frequencies, *Statistics and Probability Letters.,* **26** (1996), 51–60.

[45] J. N. Srivastava and B. C. Gupta, Main effect plan for 2^m factorials which allow search and estimation of one unknown effect, *J. Statist. Plann. Inference,* **3** (1979), 259–265.

[46] J. N. Srivastava and Y. C. Wang, Row-column designs: non-additivity makes them hazardous to use, *J. Statist. Plann. Inference,* **73** (1998), 277–315.

S. Ghosh
University of California
Riverside
USA
subir.ghosh@ucr.edu

T. Shirakura
Kobe University
Japan
sirakura@kobe-u.ac.jp

J. N. Srivastava
Colorado State University
Fort Collins
USA
jsrivas@lamar.colostate.edu

BOLYAI SOCIETY MATHEMATICAL STUDIES, 16

Entropy, Search, Complexity, pp. 113–150.

Information Topologies with Applications

P. HARREMOËS*

Topologies related to information divergence are introduced. The conditional limit theorem is taken as motivating example, and simplified proofs of the relevant theorems are given. Continuity properties of entropy and information divergence are discussed.

1. Introduction and preliminaries

Relating results from probability theory and information theory is not a new idea. Some convergence theorems in probability theory can be reformulated as "the entropy converges to its maximum". Markov chains were treated by A. Rényi [26], D. G. Kendall [20] and J. Fritz [14], the Central Limit Theorem was treated by Linnik [23] and A. Barron [2], the local central limit theorem was treated by S. Takano [27] and Poisson's law was treated by P. Harremoës [16]. There has also been work strengthening weak or strong convergence to convergence in information divergence. All the above mentioned papers have results of this kind, but also a work by A. Barron [3] should be mentioned. Some work has also been done where the limit of a sequence is identified as a information projection. The most important paper in this direction is [8], but the subject is difficult in the sense that one has to be very careful with regularity conditions. A. Dembo and O. Zeitouni discussed large deviations in [10], but here the technique was more convex

*The research is supported by a post-doc fellowship from the Villum Kann Rasmussen Foundation and by IMPAN-BC, the Banach Center, Warsaw and INTAS (project 00-738) and Danish Natural Science Counsil.

analysis than information theory; this only emphasizes the close relationship between convex analysis and information theory.

The term "convergence in information" has been used in the literature to mean convergence to zero of information divergence. In this paper it will be shown how to associate not one but at least two topologies capturing the concept of converge in information. One of these was recognized by R. Dudley but was not studied in detail [11]. The information topologies will make it easier to formulate some theorems in a simpler way than found in the literature. The topological problems are not equally important for all applications, but when discussing the conditional limit theorem and the Sanov property they are central.

The rest of this section is devoted to preliminaries on the minimum information game and to some results in topology. In Section 2 a new topology, the strong information topology, is defined and its basic properties are studied. The continuity of the entropy function and information divergence is studied in Section 3. In Section 4 the weak information topology is introduced and characterized. In Section 5 the information topologies are used in stating and proving results related to conditional limit theorem and Sanov property. The paper ends with a short discussion.

1.1. The minimum information game

For a more throughout discussion of the game theoretic approach to divergence minimization one should consult the article [28] of F. Topsøe. Most of the definitions and notation is taken from this article, but there are some exceptions which will now be described.

Let $M_+^1(U)$ denote the set of probability measures on U equipped with a σ-algebra, and let M_Q denote the set of probability measures absolutely continuous with respect to Q. The *information divergence D from P to Q* is defined by

$$D(P \parallel Q) = \begin{cases} \displaystyle\int \log \frac{dP}{dQ}\, dP, & \text{for } P \in M_Q \\ \infty, & \text{otherwise.} \end{cases}$$

where Q, P are probability measures. The information divergence is nonnegative and lower semi continuous as function of P and Q when $M_+^1(U)$ is equipped with the topology τ. In the literature information divergence is sometimes termed relative entropy and denoted $H(P,Q)$ or $S(P,Q)$. This has created a lot of confusion because information divergence has the opposite sign than entropy. Information divergence is also called Kullback–Leibler discrimination information and denoted $I(P,Q)$. Unfortunately this notation is easily confused with the notation of mutual information and therefore we will follow I. Csiszár and use the notation $D(P \parallel Q)$.

We shall use Pinsker's inequality

$$\frac{1}{2}\|P-Q\|^2 \leq D(P \parallel Q),$$

where the *total variation norm* is defined by

$$\|P-Q\| = \sup_{|f|\leq 1}\left\{\int_U f\,dP - \int_U f\,dQ\right\}.$$

If $C \subseteq M_+^1(U)$ and Q is a probability measure then *the divergence from C to Q* is defined by

$$D(C \parallel Q) = \inf_{P\in C} D(P \parallel Q).$$

The following theorem is due to I. Csiszár [7] and F. Topsøe [29].

Theorem 1. *Let $C \subseteq M_Q$ be convex and assume that $D(C \parallel Q) < \infty$. Then there exists a unique distribution $Q_{|C}$ such that $Q_n \to Q_{|C}$ in total variation for every sequence $(Q_n) \subseteq C$ such that $D(Q_n \parallel Q) \to D(C \parallel Q)$. Furthermore, for every $P \in C$, we have*

$$D(P \parallel Q) \geq D\big(P \parallel Q_{|C}\big) + D(C \parallel Q). \tag{1}$$

A sequence $(P_n) \subseteq C$ is said to be *asymptotic optimal* if $D(P_n \parallel Q) \to D(C \parallel Q)$. The probability distribution $Q_{|C}$ is called the *generalized information projection*, and the theorem implies that for any asymptotic optimal sequence $(P_n) \subseteq C$ we have $D\big(P_n \parallel Q_{|C}\big) \to 0$. By lower semi continuity of the information divergence we have

$$D\big(Q_{|C} \parallel Q\big) \leq D(C \parallel Q),$$

Especially we see that $Q_{|C}$ is absolutely continuous with respect to Q. If $Q_{|C} \in C$ then it will be called the *information projection* [7]. An alternative term is the *relative centre of attraction* which has been used in [29]. We will often use $\langle f, P\rangle$ to denote the expectation of f with respect to P when defined. With this notation the duality between measurable functions and probability measures is emphasized.

Let U be a measurable space and C an arbitrary subset of $M_+^1(U)$. Measures in C will be called consistent. Let Q be a fixed reference probability measure in $M_+^1(U)$. A *code improvement* is a measurable function $\Delta : U \to [-\infty; \infty]$ such that the following inequality is satisfied

$$\int \exp(\Delta)\, dQ \leq 1. \tag{2}$$

Note that this is weaker than that in [28] where equality is required. If U is discrete and $\kappa(u) = -\log\left(Q(u)\right)$ and $\tilde{\kappa}(u) = \kappa(u) - \Delta(u)$ then (2) is equivalent with *Kraft's inequality*

$$\sum_{u \in U} \exp\left(\tilde{\kappa}(u)\right) \leq 1.$$

Therefore (2) implies that $\Delta(u) = \kappa(u) - \tilde{\kappa}(u)$ can be interpreted as how much shorter a code word of u will be by replacing the code of length $\kappa(u)$ be a code word of length $\tilde{\kappa}(u)$.

Consider the two-persons zero-sum game with objective function

$$\langle \Delta, P\rangle$$

where Δ is a code improvement and P is a consistent measure, i.e. $P \in C$. If Δ is a version of $\log \frac{dR}{dQ}$ and $D(R \parallel Q) < \infty$ then $\langle \Delta, P\rangle = D(P \parallel Q \rightsquigarrow R)$ as defined in [28], and

$$D(P \parallel Q) = \sup_{\Delta} \langle \Delta, P\rangle.$$

Define the *compression* $\Gamma(\Delta)$ of the code improvement by

$$\Gamma(\Delta) = \inf_{P \in C} \langle \Delta, P\rangle.$$

Further we introduce the notation

$$D(C \parallel Q) = \inf_{P \in C} D(P \parallel Q)$$

$$\Gamma(C, Q) = \sup_{\Delta} \Gamma(\Delta).$$

The quantity $D(C \parallel Q)$ is the *divergence of the set* C and $\Gamma(C, Q)$ is the *redundancy* of Q. Then

$$\Gamma(C, Q) \leq D(C \parallel Q)$$

and the game is *in equilibrium* if $D(C \parallel Q) < \infty$ and the inequality holds with equality. If the game is in equilibrium the set C is said to be *in equilibrium* with Q. In the theory of large deviations the game theoretic equilibrium is sometimes stated as the existence of a dominating point as defined by P. Ney [25].

We shall need the following theorem is due to F. Topsøe [29, Theorem 9].

Theorem 2. *Let Q be a probability measure and let C be a convex set of measures such that $D(P \parallel Q) < \infty$ for all $P \in C$. Then the game is in equilibrium.*

The Radon–Nikodym derivative $\frac{dQ_{|C}}{dQ}$ is uniquely defined only as an element in $L^1(U, Q)$. Remark that if the underlying space is countable and all elements in C are distributions with finite support then the condition $D(P \parallel Q) < \infty$ is satisfied if the support of P is a subset of the support of Q.

Not all convex sets are in equilibrium with a given distribution Q. On a infinite set U a nontrivial example is the set

$$C = \left\{ P \in M_+^1(U) \mid D(P \parallel Q) = \infty \right\}.$$

For this set $D(C \parallel Q) = \infty$ and $\Gamma(C, Q) = 0$.

The most important example of information minimization is associated with conditional probability to which little attention has been paid in the literature – perhaps because it is almost trivial.

Example 3 (Conditional probability). Let Q be a distribution and let K be a subset of U with $Q(K) > 0$. Define $C_K = \left\{ P \mid P(K) = 1 \right\}$. First we observe that $Q(\cdot \mid K) \in C_K$. If we choose $\frac{dQ(\cdot|K)}{dQ} = \frac{1}{Q(K)} \cdot 1_K$ as version of the Radon–Nikodym derivative, then

$$\left\langle \log \left(\frac{dQ(\cdot \mid K)}{dQ} \right), P \right\rangle = \log \frac{1}{Q(K)}$$

for every $P \in C_K$. Therefore $Q(\cdot \mid K)$ is the minimum information distribution in C_A and

$$D(C_K \parallel Q) = \log \frac{1}{Q(K)}.$$

Further $\log \frac{dQ(\cdot|K)}{dQ} = D(C_K \parallel Q)$ almost $Q(\cdot \mid K)$-surely.

Note that the choice of version is important in this example if $Q(K) < 1$.

The principle of minimum information discrimination as formulated by S. Kullback in [22, Chapter 5] states that when an empirical distribution P is given then the hypothesis Q minimizing $D(P \parallel Q)$ should be chosen. This is a reformulation of the maximum likelihood idea known from statistics. The principle extends to the situation when a prior distribution Q is given and and P is unknown and some additional information is provided (the set C). Then the principle states that one shall replace the original distribution Q by that consistent distribution P which minimizes the information divergence $D(P \parallel Q)$. On a finite set with a uniform prior the extended principle is equivalent to Jaynes' Maximum Entropy Principle [18]. The example demonstrated that conditional probability fits into this general principle. The conditional limit theorem extends the example to more general sets of consistent distributions.

1.2. Topologies on the set of probability measures

On the set of probability distributions $M_+^1(U)$ there exist several relevant topologies. The τ-topology is a weak topology defined as the initial topology corresponding to the maps

$$P \curvearrowright \int_U f\,dP,$$

where f is required to be measurable and bounded.

The *strong topology* is defined by the total variation norm. On a countable set the strong and the τ-topology coincide.

In [8] I. Csiszár introduced a topology τ_0 somewhat stronger than the τ-topology. This is defined by the basic neighborhoods

$$\bigl\{P \in M_+^1(U) \bigm| \left|P(B_j) - Q(B_j)\right| < \varepsilon,$$

$$P(B_j) = 0 \text{ if } Q(B_j) = 0,\ j = 1, 2, \ldots, k\bigr\},$$

where $B_1, B_2, \ldots, B_k$ is a measurable partition of U. For all sequences $(P_n)_{n\in\mathbb{N}}$ the sequence P_n converge to Q in τ_0, if and only if $P_n \xrightarrow{\tau} Q$ and $P_\lambda \preccurlyeq Q$ eventually, where $\preccurlyeq$ means absolutely continuous with respect to.

Proposition 4. *The distributions P and Q belong to the same τ_0-connected component if and only if P and Q are equivalent, that is mutual absolutely continuous.*

Proof. Assume that P and Q are not equivalent. Without loss of generality we may assume that $Q \not\ll P$. Then there exists B such that $Q(B) > 0$ and $P(B) = 0$. Let C denote the set $\{R \in M_+^1(U) \mid R(B) = 0\}$. Then C is open and closed and $P \in C$ and $Q \in \complement C$. Therefore P and Q belong to complementary open sets and therefore to different connected components.

Assume that P and Q are equivalent. Then $(1-t)P + tQ$, $t \in [0;1]$ is a continuous curve from P to Q and therefore P and Q belong to the same connected component. ■

If U is finite, then the τ_0-connected components are the probability distributions with given support, and on each of these connected components τ_0 and τ are equal. Other topologies stronger than τ has also been proposed [12] for clarification of conditions for the conditional limit theorem to hold.

Since long it has been known that the divergence balls

$$B(Q,r) = \{P \in M_+^1(U) \mid D(P \parallel Q) \leq r\}$$

do not define a basis of a neighborhood filter of a topology, see [5], [6] and [11]. However, the divergence *does* define several topologies relevant to information theory. Their relation is very similar to the relation between the strong topology and the τ-topology, but they are somewhat stronger than these topologies.

Here we shall recall some useful facts from general topology, see [13] and [21] for more details.

Definition 5. An $\mathcal{L}^*$-space is a pair $(X, \mathcal{L})$ where X is a set and $\mathcal{L}$ is a function (called *the limit operator*) assigning to some sequences of points *of* X an element *of* X (called the *limit of the sequence*) in such a way that the following conditions are satisfied (we write $\mathcal{L}x_n$ instead of $\mathcal{L}(\{x_n\})$ and say that $\{x_n\}$ *converges* to x if $\mathcal{L}x_n = x$):

L1 If $x_n = x$ for $n = 1, 2, \ldots$, then $\mathcal{L}x_n = x$.

L2 If $\mathcal{L}x_n = x$ then $\mathcal{L}x_{k_n} = x$ for every subsequence $\{x_{k_n}\}$ of $\{x_n\}$.

L3 If a sequence $\{x_n\}$ does not converge to x, then it contains a subsequence $\{x_{k_n}\}$ such that no subsequence of $\{x_{k_n}\}$ converges to x.

The sequential closure $\mathrm{cl}_\sigma(A)$ of a set $A \subseteq X$ is the set of elements $x \in X$ such that there exists a sequence $\{x_n\}$ in X such that $\mathcal{L}x_n = x$.

Theorem 6. *Let $(X, \mathcal{L})$ be an $\mathcal{L}^*$-space. Then there exists a topology (called the sequential topology) on X such that A is closed in the sequential topology if and only if $A = \mathrm{cl}_\sigma(A)$. A sequence $\{x_n\}$ converges to x in the sequential topology if and only if $\mathcal{L}x_n = x$.*

2. The strong information topology

We shall start by defining the new topology on the set of probability measures. A sequence P_n of probability measures is said *to converges to Q strongly in information* if $D(P_n \parallel Q) \to 0$. It is easy to check that the conditions L1-L3 are satisfied. Therefore an $\mathcal{L}^*$-space is defined and the corresponding sequential topology is called *the strong information topology* and is denoted I. This topology was first recognized by R. Dudley [11, Thm. 3.8], but has not been studied in any detail in the literature.

Definition 7. A subset $A \subseteq M_+^1(U)$ is said to be I-closed if $D(A \parallel Q) = 0$ imply $Q \in A$.

A set A is open in the strong information topology if and only it satisfies one of the following equivalent conditions:

- For any $Q \in A$ there exists a $r > 0$ such that $B(Q, r) \subseteq A$.
- For any net $(P_\lambda)_{\lambda \in \Lambda}$ in $M_+^1(U)$ which satisfies $D(P_\lambda \parallel Q) \to 0$ for some $Q \in A$, there exists a λ_0 such that $P_\lambda \in A$ for $\lambda \geq \lambda_0$.
- For any sequence $(P_n)_{n \in \mathbb{N}}$ in $M_+^1(U)$ which satisfies $D(P_n \parallel Q) \to 0$ for some $Q \in A$, there exists a n_0 such that $P_n \in A$ for $n \geq n_0$.

Note that the strong information topology can be defined directly by the above characterization of the I-closed/open sets.

Theorem 8. *The strong information topology is stronger than the topologies defined by total variation and τ_0.*

Proof. We have to show that the balls $\left\{ P \in M_+^1(U) \mid \|P - Q\| < r \right\}$ are open in the information topology. Assume that

$$P_0 \in \left\{ P \in M_+^1(U) \mid \|P - Q\| < r \right\},$$

and that $(P_n)_{n\in\mathbb{N}}$ is a sequence such that $D(P_n \parallel P_0) \to 0$. Then by Pinsker's inequality $\|P_n - P_0\| \to 0$, and there exists a n_0 such that $\|P_n - P_0\| < r - \|P_0 - Q\|$ for $n \geq n_0$. Therefore

$$\|P_n - Q\| \leq \underbrace{\|P_n - P_0\| + \|P_0 - Q\|}_{<r}$$

and $P_n \in \{P \in M_+^1(U) \mid \|P - Q\| < r\}$ for $n \geq n_0$.

In order to see the I is stronger than τ_0 we just have to remark that for any measurable set $B \subseteq U$ the set

$$\{P \in M_+^1(U) \mid P(B) = 0\}$$

is I-open. ■

In general the strong information topology coincides with neither τ_0 nor the strong topology. To see this consider a countable set U with equivalent probability measures P and Q such that $D(P \parallel Q) = \infty$. Put $P_n = 1/n \cdot P + (1 - 1/n) \cdot Q$. Then $P_n \to Q$ in strong topology and in τ_0 but not in the strong information topology.

Lemma 9. *A one parameter exponential family is continuous in the parameter in the strong information topology.*

Proof. Let P_t be a one parameter exponential family for which $P_0 = P$ and $P_1 = Q$. First we remark that $D(P_s \parallel P_t)$ can only be infinite for $s = 0$ and $t = 1$ or $s = 1$ and $t = 0$. For $t' \in [-D(P \parallel Q); D(Q \parallel P)]$ let $Q_{t'}$ denote the distribution P_t such that

$$t' = \int \log \frac{dQ}{dP}\, dP_t.$$

This is a new parametrization of the family, and the change in variable is continuous. We shall show that $t \curvearrowright Q_t$ is continuous when $M_+^1(U)$ is equipped with the strong information topology. For a fixed value of t_0 the function

$$t \curvearrowright D(Q_t \parallel Q_{t_0})$$

is convex because Q_s minimizes $D(R \parallel Q_{t_0})$ under the constraint $\int \log \frac{dQ}{dP}\, dR$ and $D(Q_{t_0} \parallel Q_{t_0}) = 0$. Therefore $t \curvearrowright D(Q_t \parallel Q_{t_0})$ is continuous in $t = t_0$ because $D(Q_t \parallel Q_{t_0}) < \infty$ for t in a neighborhood of t_0. ■

Theorem 10. *The distributions P and Q belong to the same connected component if and only if P and Q are equivalent.*

Proof. Assume that P and Q are not absolutely continuous with respect to each other. Then P and Q belong to different τ_0-components and therefore also to different I-components.

Assume that P and Q are equivalent. Now we shall define a continuous path from P to Q. For $t \in \{0, 1\}$ the integral

$$\int \left(\frac{dQ}{dP}\right)^t dP$$

equals one. The function

$$t \curvearrowright \int \left(\frac{dQ}{dP}\right)^t dP$$

is convex and therefore

$$\int \left(\frac{dQ}{dP}\right)^t dP < 1$$

for $t \in]0; 1[$. Using that P and Q are equivalent we see that the integral is positive. Then a log-affine curve $t \curvearrowright P_t$ is given by the equation

$$\frac{dP_t}{dP} = \frac{\left(\frac{dQ}{dP}\right)^t}{\int \left(\frac{dQ}{dP}\right)^t dP}.$$

Therefore P and Q are connected by the log-affine curve. ∎

In the proof the one-parameter exponential family from P to Q was used. This family is closely related to the so-called Chernoff bound known from statistics [4, section 12.9]. It can also be considered as a geodetic curve when the set of probability measures are equipped with a suitable geometric structure related to information divergence [1].

Theorem 11. *If U is finite then the strong information topology coincides with the topology τ_0.*

Proof. On a finite set $D(P \parallel Q) < \infty$ if and only if $\mathrm{supp}\,(P) \subseteq \mathrm{supp}\,(Q)$. The previous theorem shows that the connected components are of the form $\left\{P \in M_+^1(U) \mid \mathrm{supp}\,(P) = B\right\}$. Now we just have to remark that

$$(P, Q) \curvearrowright D(P \parallel Q)$$

is τ_0-continuous on each component. ∎

Corollary 12. *The net $P_\lambda \in M_+^1(U)$ converges to $Q \in M_+^1(U)$ in the topology τ_0 if and only if $\Phi(P_\lambda) \to \Phi(Q)$ in the strong information topology for all finite partitions Φ, i.e. all $\Phi : U \to V_\Phi$ where V_Φ is a finite set.*

As immediate consequences of Theorem 8 one sees that the map $(P, Q) \curvearrowright D(P \parallel Q)$ is lower semi continuous when $M_+^1(U)$ is equipped with the strong information topology. Especially the sets

$$\{P \in M_+^1(U) \mid D(P \parallel Q) \leq r\}$$

are closed in the strong information topology. In general the sets

$$\{P \in M_+^1(U) \mid D(P \parallel Q) < r\}$$

are not open.

Example 13. If U is countable infinite then there exists a ball $B(Q, r)$ such that Q is not an I-interior point. To see this choose P, R, Q with point probabilities proportional to $\frac{1}{k^2}, \frac{1}{k^3}, \frac{1}{2^k}$. Then

$$D(P \parallel R) < \infty$$
$$D(R \parallel Q) < \infty$$
$$D(P \parallel Q) = \infty.$$

Define $P_n = \frac{1}{n}P + \left(1 - \frac{1}{n}\right)R$. Then

$$D(P_n \parallel R) \leq \frac{1}{n} D(P \parallel R) + \left(1 - \frac{1}{n}\right) D(R \parallel R)$$
$$= \frac{1}{n} D(P \parallel R) \to 0.$$

Therefore $P_n \xrightarrow{I} R$ for $n \to \infty$ and $D(P_n \parallel Q) = \infty$.

The above construction can be made with P replaced by $P_\alpha = \alpha P + (1 - \alpha)Q$ and R replaced by $R_\alpha = \alpha R + (1 - \alpha)Q$ for $\alpha \in]0; 1[$. Therefore $\alpha R + (1 - \alpha)Q$ is on the I-boundary. We have

$$D(R_a \parallel Q) = D\big(\alpha R + (1 - \alpha)Q \parallel Q\big)$$
$$\leq \alpha D(R \parallel Q) \to 0 \quad \text{for} \quad \alpha \to 0.$$

This shows that there exists a net $P_\lambda \to Q$ such that $D(P_\lambda \parallel Q) = \infty$. Similar constructions are known from the literature, cf. [6] and [8, Example 3.2].

A set A is closed if and only if $\mathrm{cl}_\sigma(A) = A$. In general it is not true that $\mathrm{cl}_\sigma(A)$ is closed. Therefore we have to iterate the sequential closure operation to obtain the closure $\mathrm{cl}_I(A)$. The construction works as follows:

Let α be an ordinal (finite or transfinite). We define $\mathrm{cl}_\sigma^\alpha(A)$ by transfinite recursion:

- $\mathrm{cl}_\sigma^\alpha(A) = A$ for $\alpha = 0$
- $\mathrm{cl}_\sigma^\alpha(A) = \mathrm{cl}_\sigma\Big(\bigcup_{\beta<\alpha} \mathrm{cl}_\sigma^\beta(A)\Big)$ for $\alpha > 0$.

We see that $\mathrm{cl}_\sigma^\alpha(A)$ is constant from a certain point. If not before then when the cardinality of α exceeds the cardinality of U. Therefore $\mathrm{cl}_\sigma^\alpha(A)$ is closed. A more refined argument gives that $\mathrm{cl}_\sigma^{\omega_1}(A)$ is closed and equals $\mathrm{cl}_I(A)$ where ω_1 denote the first non-countable ordinal number.

Let L denote the convex hull of P, Q and R. Then $L = \mathrm{cl}_\sigma^2\big(\mathrm{int}\,(L)\big)$ where $\mathrm{int}\,(L)$ denotes the algebraic interior of L. The above construction shows that 2 closure operations are needed in order to obtain a closed set. This construction can be iterated in order to obtain a set for which n closure operations are needed in order to obtain a closed set. The construction is as follows. Let P_k be the probability measure

$$\bigotimes^k Q \otimes R \otimes \bigotimes^{n-k} P$$

for $k = 0, 1, \ldots, n$. Then

$$\begin{aligned} D(P_k \parallel P_{k+1}) &= D(P \parallel R) + D(R \parallel Q) \\ &< \infty. \end{aligned}$$

In general

$$D(P_k \parallel P_l) < \infty$$

if and only if

$$l \leq k + 1.$$

For probability vectors (α_k) and (β_k) we have

$$D\Big(\sum_{k=1}^n \alpha_k P_k \,\Big\|\, \sum_{k=1}^n \beta_k P_k\Big) < \infty$$

if and only

$$\min_{\beta_k>0} k \leq \min_{\alpha_k>0} k + 1.$$

Let C be the convex hull of $\{P_k\}$ and let K be the algebraic interior of C. Then $\mathrm{cl}_\sigma^k(K) = C \backslash conv_{l>k}\{P_l\}$. Therefore n closure operations are needed to obtain a closed set.

Theorem 14. *Let A be a convex subset of $M_+^1(U)$. Then $\mathrm{cl}_I(A)$ is a convex set.*

Proof. Let $\mathrm{cl}_\sigma(B)$ denote the sequential closure of B, i.e. set of distributions Q such that there exists a sequence $(P_n)_{n\in\mathbb{N}}$ such that $P_n \xrightarrow{I} Q$ and $P_n \in B$. If B is convex then so is $\mathrm{cl}_I^\sigma(B)$. To see this assume that $P_n \xrightarrow{I} P$ and $Q_n \xrightarrow{I} Q$ where $P_n, Q_n \in B$. Let $\alpha \in [0;1]$. Then

$$D\big(\alpha P_n + (1-\alpha)Q_n \parallel \alpha P + (1-\alpha)Q\big)$$

$$\leq \alpha D(P_n \parallel P) + (1-\alpha)D(Q_n \parallel Q) \to 0.$$

By transfinite induction we see that also the closure is convex. ■

A non-trivial example of a convex set which is closed in the strong information topology is the set of atomic probability measures. To see that this set is closed let P_n be a sequence of atomic probability measures which converges to Q in strong information topology. Let A_n denote the set of points where P_n has positive weight. Then A_n is countable. Thus all P_n are concentrated on the countable set $\bigcup A_n$, and therefore Q is also concentrated on $\bigcup A_n$. Therefore Q is atomic.

We easily see that the strong information topology is compact if and only if U is a singleton.

Proposition 15. *For the strong information topology the following conditions are equivalent.*

1. *U is finite.*
2. *$M_+^1(U)$ is $\sigma-$compact.*
3. *$M_+^1(U)$ is locally compact.*
4. *1st countability axiom is satisfied.*
5. *2nd countability axiom is satisfied.*

Proof. 1. $\Rightarrow$ 2. Obvious.

2. $\Rightarrow$ 3. Obvious.

$\neg 1. \Rightarrow \neg 3.$ Assume that U is not finite, and let Q be a probability measure with infinite but countable support. Assume that ω is a compact neighborhood in strong information topology. Then ω contains a ball $B(Q,r) \subseteq \omega$, and the balls are closed so $B(Q,r)$ is compact. Now there exists a sequence $P_n \to Q$ in I such that P_n has finite support. Therefore $B(Q,r)$ contains a probability distribution P with finite support. The sequence $\left(1-\frac{1}{n}\right)P+\frac{1}{n}Q \to P$ in weak topology but not in strong information topology. Therefore the sequence $\left(1-\frac{1}{n}\right)P+\frac{1}{n}Q$ has no convergent subnet, and the contradiction is obtained.

1. $\Rightarrow$ 5. Obvious.

5. $\Rightarrow$ 4. Obvious.

$\neg 1. \Rightarrow \neg 4.$

Assume that U is not finite. Then there exists a convex set C such that $\mathrm{cl}_\sigma\left(\mathrm{cl}_\sigma(C)\right) \neq \mathrm{cl}_\sigma(C)$ where the sequential closure is taken in the relevant information topology. Therefore the strong information topology does not satisfy 1st countability axiom. ■

The conditions in Proposition 15 imply paracompactness, but it is not clear weather they are necessary conditions.

Proposition 16. *The strong information topology is separable if and only if U is countable.*

Proof. If U is finite the strong information topology is obviously separable.

If U is countable then we may assume that $U = \mathbb{N}$. Put $A_n = \{1, 2, \ldots, n\}$ and $C_n = \left\{ P \mid \mathrm{supp}\,(P) \subseteq A_n \right\}$. Each of the sets C_n is separable, and therefore their union $\bigcup C_n$ is separable. Now

$$D(C_n \parallel Q) = D\left(Q_{|C_n} \parallel Q\right)$$
$$= \log \frac{1}{Q(A_n)} \to 0.$$

Therefore $\bigcup C_n$ is dense in $M_+^1(U)$, and $M_+^1(U)$ is separable in the strong information topology.

If U is not countable, then $M_+^1(U)$ is not separable in τ and therefore not in the strong information topology. ■

3. Continuity of entropy and divergence

In order to characterize the sequential discontinuity points of the entropy and divergence functions we need the concept of hyperbolic distribution and several ways to characterize them.

Definition 17. A distribution on $\mathbb{R}$ with distribution function F is said to *tail majorize* a distribution with distribution G if

$$F(x) \geq G(x)$$

eventually, i.e. there exists a x_0 such that $F(x) \geq G(x)$ for $x \geq x_0$. A distribution of a non-negative random variable with distribution function F is *power dominated* if there exists $\alpha > 1$ such that F is tail majorized by the distribution with probability density proportional to $x^{-\alpha}$ for $x \geq 1$. A distribution of a non-negative random variable is said to be hyperbolic if it is not power dominated.

Proposition 18. *Let X be a non-negative random variable with distribution function F. Then the following conditions are equivalent:*

Lemma 19.

1. *F is power dominated.*
2. *There exists a $t > 0$ such that*

$$E(X^t) < \infty.$$

3. *The tail probabilities satisfy*

$$\liminf_{x \to \infty} \frac{\log\left(1 - F(x)\right)}{\log \frac{1}{x}} > 0.$$

Proof. 1. $\Rightarrow$ 2. Assume that F majorizes a power law G. Then

$$\int x^t \, dF(x) \leq \int x^t \, dG(x),$$

because the function $x \to x^t$ is increasing. The right hand side is finite for t sufficiently small.

2. ⇒ 3. For all t we have

$$E(X^t) = \int_0^\infty x^t \, dF(x)$$
$$= \sum_{i=0}^{\infty} \int_i^{i+1} x^t \, dF(x)$$
$$\geq \sum_{i=0}^{\infty} i^t \; P(i \leq X < i+1).$$

The series $\sum_{i=0}^{\infty} i^{-s} \, P(i \leq X < i+1)$ is an ordinary Dirichlet series in $s = -t$ [15] which converges for $t = 0$. Thus the abscissa of convergence is

$$\gamma = \limsup_{n \to \infty} \frac{\log \left| 1 - F(n) \right|}{\log(n)}.$$

If $E(X^t) < \infty$ then

$$\limsup_{n \to \infty} \frac{\log \left| 1 - F(n) \right|}{\log(n)} \leq -t < 0.$$

3. ⇒ 1. Assume that

$$\liminf_{x \to \infty} \frac{\log \left(1 - F(x)\right)}{\log \frac{1}{x}} > 0.$$

Then there exists $t > 0$ such that

$$\frac{\log \left(1 - F(x)\right)}{\log \frac{1}{x}} \geq t$$

eventually, but this is equivalent to

$$1 - F(x) \leq x^t$$

eventually, and we see that F is power dominated. ■

Typical examples of hyperbolic distributions are distributions with densities proportional to

$$\frac{1}{x(\log x)^{\alpha}}, \qquad x \geq 2$$

for some $\alpha > 1$.

For a random variable X with probability density f the Rényi entropy of order $\alpha > 0$ and $\alpha \neq 1$ is defined by the formula

$$h^{\alpha}(P) = \frac{1}{1-\alpha} \log \int \left(f(x)\right)^{\alpha} dx$$

when the integral converges. For $\alpha = 1$ the Rényi entropy $h^1(P)$ is defined to be the Shannon entropy (differential entropy). Note that h is used to denote differential entropy and H is used to denote discrete entropy.

For continuous distributed non-negative random variables with decreasing density further characterization of power dominated and hyperbolic distributions are possible.

Proposition 20. *Let X be a continuous non-negative random variable with distribution function F and density function $f = F'$. Assume that f upper bounded and decreasing. Then the following conditions are equivalent:*

1. *The distribution of X is power dominated.*
2. *There exists $K > 0$ and $\alpha > 1$ such that $f(x) \leq K \cdot x^{-\alpha}$ eventually.*
3. *There exists $\alpha > 1$ such that for all $L > 0$ one has $f(x) \leq L \cdot x^{-\alpha}$ eventually.*
4. *The density function satisfies*

$$\liminf \frac{\log f(x)}{\log \frac{1}{x}} > 1.$$

5. *There exists $\alpha < 1$ such that $H^{\alpha}(X) < \infty$.*

Proof. 2. $\Rightarrow$ 3. Assume that $K > 0$ and $\alpha > 1$ such that $f(x) \leq K \cdot x^{-\alpha}$ eventually. Choose β such that $1 < \beta < \alpha$. Then for any $L > 0$

$$K \cdot x^{-\alpha} \leq L \cdot x^{-\beta}.$$

3. $\Rightarrow$ 1. This follows since the constant L can be chosen such $L \cdot x^{-\alpha}$, $x \geq 1$ is a power law.

4. $\Rightarrow$ 2. If

$$\liminf \frac{\log f(x)}{\log \frac{1}{x}} > \alpha > 1$$

then

$$\frac{\log f(x)}{\log \frac{1}{x}} > \alpha$$

eventually and this is equivalent to $f(x) < x^{-\alpha}$ eventually.

5. ⇒ 4. Assume that the Rényi entropy $H^{\alpha}(P)$ is finite for $\alpha < 1$. Then

$$\begin{aligned}\sum_{n=1}^{\infty} \exp\big(-\alpha(-\log f(n))\big) &= \sum_{n=1}^{\infty} \left(f(n)\right)^{\alpha} \\ &\leq \int_0^{\infty} \left(f(x)\right)^{\alpha} dx \\ &< \infty.\end{aligned}$$

Now $\sum_{n=1}^{\infty} \exp\big(-\alpha(-\log f(n))\big)$ is a Dirichlet series in $-\alpha$ [15] with abscissa of convergence equal to

$$\limsup_{n\to\infty} \frac{\log n}{-\log f(n)} = \frac{1}{\liminf_{n\to\infty} \frac{\log f(n)}{\log \frac{1}{n}}}.$$

Therefore

$$\liminf \frac{\log f(x)}{\log \frac{1}{x}} \geq \frac{1}{\alpha}.$$

1. ⇒ 5. Assume that the power law with tail probabilities $1 - G(x) = x^{-\alpha}$, $x \geq 1$ is majorized by F. The function $x \to x^{\beta}$, $\beta < 1$ is concave. Using Shur convexity [24] we have

$$\sum_{j=0}^{\infty} \left(F'\left(\frac{j}{n}\right)\right)^{\beta} \leq \sum_{j=0}^{\infty} \left(G'\left(\frac{j}{n}\right)\right)^{\beta},$$

and

$$\int_0^{\infty} \left(F'(x)\right)^{\beta} dx \leq \int_0^{\infty} \left(G'(x)\right)^{\beta} dx$$

The right hand side is finite for β sufficiently close to 1. ■

A non-negative random variable with decreasing density f always satisfies

$$f(x) \leq \frac{1}{x}$$

with equality in case the distribution is uniform. Therefore

$$\frac{\log f(x)}{\log \frac{1}{x}} \geq 1.$$

Therefore a hyperbolic distribution with decreasing density satisfies

$$\liminf_{x\to\infty} \frac{\log f(x)}{\log \frac{1}{x}} = 1.$$

The (discrete) Rényi entropy of order $\alpha > 0$ and $\alpha \neq 1$ is defined as a extended real number by the formula

$$H^{\alpha}(P) = \frac{1}{1-\alpha} \log \sum p_i^{\alpha}$$

For $\alpha = 1$ the Rényi entropy $H^1(P)$ is defined to be the Shannon entropy. Then Rényi entropy is decreasing in α.

Assume that the random variable X is non-negative and integer valued and with decreasing point probabilities. Then the previous result can be applied to the distribution with density f given by

$$f(x) = P\big(X = \lfloor x \rfloor\big).$$

Then the discrete (Shannon or Rényi) entropy of X equals the differential entropy of f. The distribution of X is hyperbolic if and only if

$$\liminf_{i\to\infty} \frac{\log P(X=i)}{\log \frac{1}{i}} = 1.$$

This formula was used to characterize discrete hyperbolic distributions in [17].

After these characterizations of the hyperbolic distributions it is possible to study the sequential discontinuity points of entropy and divergence. Let Q be a discrete distribution. If Q is power bounded then the entropy is finite because the entropy of a power law is finite. The following theorem identifies the hyperbolic distributions as the (sequential) discontinuity points of the entropy function.

Theorem 21. *If Q is power dominated, then $P_n \xrightarrow{I} Q$ implies $H(P_n) \to H(Q)$. If Q is hyperbolic and $H(Q) < \infty$ then there exists a sequence P_n such that $P_n \xrightarrow{I} Q$ and*

$$\liminf_{n\to\infty} H(P_n) > H(Q).$$

Proof. Assume that Q is power bounded. Then

$$\sum_i \exp\left(t \log\left(\frac{1}{q_i}\right)\right) q_i = \sum_i q_i^{1-t} < \infty$$

for t sufficiently small. Therefore

$$\begin{aligned} H(P_n) &= \sum_i \log\left(\frac{1}{q_i}\right) \cdot P_n(i) - D(P_n \parallel Q) \\ &\to \sum_i \log\left(\frac{1}{q_i}\right) \cdot q_i - 0 \\ &= H(Q). \end{aligned}$$

For the last part of the theorem, see [17]. ■

It is obvious to ask to what extent the function $P \curvearrowright D(P \parallel Q)$ is continuous in the information topologies. We define the Rényi divergence $D^s(P_0 \parallel Q) = \frac{1}{1-s} \log \int \left(\frac{dP_0}{dQ}\right)^s dQ$. We have $D^s(P_0 \parallel Q) \to D(P_0 \parallel Q)$ for $s \to 1_-$. If $D(P_0 \parallel Q) < \infty$, then the following conditions are equivalent.

Proposition 22.

1. $\int \left(\frac{dP_0}{dQ}\right)^t dP_0 < \infty$ *for some* $t > 0$.
2. $\int \left(\frac{dP_0}{dQ}\right)^s dQ < \infty$ *for some* $s > 1$.
3. $D^s(P_0 \parallel Q) < \infty$ *for some* $s > 1$.
4. *The random variable* $\frac{dP_0}{dQ}$ *is power dominated with respect to* P_0.

Theorem 23. *Assume that* $D(P_0 \parallel Q) < \infty$. *Then the following conditions are equivalent.*

1. *The function* $P \curvearrowright D(P \parallel Q)$ *is sequential continuous in* P_0 *in the strong information topology.*
2. *The random variable* $\frac{dP_0}{dQ}$ *is power dominated with respect to* P_0.

Proof. 2. $\Longrightarrow$ 1.: Assume that $\int \left(\frac{dP_0}{dQ}\right)^t dP_0 < \infty$ for a $t > 0$, and assume that $D(P_n \parallel P_0) \to 0$. We write

$$D(P_n \parallel Q) = D(P_n \parallel P_0) + \left\langle \log \frac{dP_0}{dQ}, P_n \right\rangle$$

We have

$$\int \exp\left(t \cdot \log \frac{dP_0}{dQ}\right) dP_0 = \int \left(\frac{dP_0}{dQ}\right)^t dP_0 < \infty.$$

Therefore

$$\left\langle \log \frac{dP_0}{dQ}, P_n \right\rangle \to \left\langle \log \frac{dP_0}{dQ}, P_0 \right\rangle$$

$$= D(P_0 \parallel Q) \quad \text{for} \quad n \to \infty.$$

$\neg 2. \implies \neg 1.$ Let b be some real number greater than $D(P_0 \parallel Q)$ and assume $\int \left(\frac{dP_0}{dQ}\right)^s dQ = \infty$ for all $s > 1$. Put

$$C_b = \left\{ P \mid \left\langle \log \frac{dP_0}{dQ}, P \right\rangle = b \right\}.$$

Then for $P \in C_b$

$$D(P \parallel Q) = \left\langle \log \frac{dP_0}{dQ}, P \right\rangle + D(P \parallel P_0)$$

$$\geq b.$$

Put $A_c = \left\{ u \mid b \leq \log \frac{dP_0}{dQ}(u) \leq c \right\}$ where c is some constant such that $Q(A_c) > 0$. Then $D\left(Q_{|A_c} \parallel Q\right) = -\log Q(A_c) < \infty$. Further

$$b \leq \left\langle \log \frac{dP_0}{dQ}, Q_{|A_c} \right\rangle \leq c.$$

Therefore there exists a convex combination $P' = \gamma Q_{|A_c} + (1-\gamma) P_0$ such that $P' \in C_b$ and

$$D(P' \parallel Q) \leq \gamma D\left(Q_{|A_c} \parallel Q\right) + (1-\gamma) D(P_0 \parallel Q)$$

$$< \infty.$$

Therefore $D(C_b \parallel Q) < \infty$. Now the projection on C_b must lie in the exponential family P^s given by

$$\frac{dP^s}{dQ} = \frac{\exp\left(s \log \frac{dP_0}{dQ}\right)}{\int \exp\left(s \log \frac{dP_0}{dQ}\right) dQ}$$

$$= \frac{\frac{dP_0}{dQ}^s}{\int \left(\frac{dP_0}{dQ}\right)^s dQ}$$

but this is not defined for $s > 1$ so the projection must be $P^1 = P_0$. ■

4. Weak information topologies

We shall define yet another topology on the set of probability measures. The relation between this and the strong information topology resembles the relation between the strong topology and τ_0.

Definition 24. A sequence $(P_n)_{n\in\mathbb{N}}$ in $M_+^1(U)$ is said to *converge to Q weakly in information* if one has $\int f\,dP_n \to \int f\,dQ$ for $n\to\infty$ for any measurable function $f : U \to [0;\infty]$ for which $\exp\big(t\cdot f(u)\big)$ is Q-integrable if $t>0$ is sufficiently small.

It is easy to check that the conditions L1-L3 are satisfied. Therefore an $\mathcal{L}^*$-space is defined and the corresponding topology is called *the sequential weak information topology.* This topology is denoted I_s^*. It is obviously a Hausdorff topology. Recall the definition of $\Gamma(C,Q)$ given by $\sup_\Delta \inf_{P\in C}\langle\Delta,P\rangle$ where the supremum is taken over all Δ satisfying $\int \exp(\Delta)\,dQ \le 1$.

Theorem 25. *Let $(P_n)_{n\in\mathbb{N}}$ be a sequence in $M_+^1(U)$ and let $Q\in M_+^1(U)$. The following 4 conditions are equivalent*

1. *For all sets C with $\Gamma(C,Q)>0$ we have $P_n\notin C$ eventually.*
2. *For all measurable functions f for which $\exp\big(t\cdot f(u)\big)$ is Q-integrable for some $t>0$, one has that $\limsup_{n\to\infty}\int f\,dP_n \le \int f\,dQ$.*
3. *P_n converges to Q in the sequential weak information topology.*
4. *$\Phi(P_n)\xrightarrow{I}\Phi(Q)$ for all measurable maps $\Phi : U\to\mathbb{N}$ such that $\Phi(Q)$ is power bounded.*

Proof. 1. $\Rightarrow$ 2.

Assume that $f : U\to[0;\infty]$ is a measurable function for which $\exp\big(t\cdot f(u)\big)$ is Q-integrable for $t\in[0;\varepsilon]$ for some $\varepsilon>0$. Put

$$\frac{dQ_t}{dQ}(u) = \frac{\exp\big(t\cdot f(u)\big)}{\int \exp\big(t\cdot f(u)\big)\,dQ(u)}.$$

Then $\int f\,dQ_t$ is continuous, strictly increasing and finite for $t\in[0;\varepsilon[$. It is easy to check that the set

$$K = \left\{P\in M_+^1(U)\,\middle|\, \int f\,dP \ge \int f\,dQ_t\right\}$$

is in equilibrium with Q and that $Q_{\varepsilon/2}$ and $\log \frac{dQ_{\varepsilon/2}}{dQ}$ are optimal strategies. Thus $\Gamma(K, Q) = D(Q_t \parallel Q) > 0$ for $t \in \,]0; \varepsilon[$. Let $(P_n)_{n\in\mathbb{N}}$ be a sequence in $M_+^1(U)$ such that for all sets C with $\Gamma(C, Q) > 0$ we have $P_n \notin C$ eventually. Then, in particular $P_n \notin K$ and $\int f\,dP_n < \int f\,dQ_t$ eventually. This holds for all $t \in \,]0; \varepsilon[$ and therefore

$$\begin{aligned}\limsup_{n\to\infty} \int f\,dP_n &\leq \liminf_{t\in]0;\varepsilon[} \int f\,dQ_t \\ &= \int f\,dQ.\end{aligned}$$

2. $\Rightarrow$ 3.

Let f be a non-negative function. Put $U_i = \{u \in U \mid i \leq f(u) < i+1\}$, $i = 0, 1, 2, \ldots$ Then using Fatou's lemma we get

$$\begin{aligned}\liminf_{n\to\infty} \int f\,dP_n &= \liminf_{n\to\infty} \int f\,dP_n \\ &= \liminf_{n\to\infty} \sum_{i=0}^{\infty} \int_{U_i} f\,dP_n \\ &\geq \sum_{i=0}^{\infty} \int_{U_i} f\,dQ \\ &= \int f\,dQ.\end{aligned}$$

This holds for all sequences P_n. If for all measurable functions f for which $\exp\big(t \cdot f(u)\big)$ is Q-integrable for some $t > 0$, one has that

$$\limsup_{n\to\infty} \int f\,dP_n \leq \int f\,dQ,$$

then $\int f\,dP_n \to \int f\,dQ$ for $n \to \infty$.

3 $\Rightarrow$ 4.

Assume that $\Phi(Q)$ is power bounded, and that (P_n) is a sequence of probability measures on the measure space $(\Omega, \mathbb{A})$ such that $P_n \xrightarrow{I_s^*} Q$. Then

$\Phi(P_n) \xrightarrow{I_s^*} \Phi(Q)$. Let p_n^i denote the point probabilities of $\Phi(P_n)$ and let q^i denote the point probabilities of $\Phi(Q)$. Then

$$D\left(\Phi(P_n) \parallel \Phi(Q)\right) = \left\langle \log \frac{d\Phi(P_n)}{d\Phi(Q)}, \Phi(P_n) \right\rangle$$

$$= \sum \log \frac{1}{q_i} \cdot p_n^i - H\left(\Phi(P_n)\right).$$

Now $\Phi(Q)$ is power bounded so there exist $t > 0$ such that $\sum_i \exp\left(t \cdot \log \frac{1}{q_i}\right) \cdot q_i = \sum q_i^{1-t} < \infty$. Therefore

$$\sum_i \log \frac{1}{q_i} \cdot p_n^i \to \sum_i \log \frac{1}{q_i} \cdot q_i$$

$$= H\left(\Phi(Q)\right)$$

$$< \infty.$$

Using lower semi continuity of the entropy function we get

$$\limsup_{n\to\infty} D\left(\Phi(P_n) \parallel \Phi(Q)\right) = \limsup_{n\to\infty} \left(\sum \log \frac{1}{q_i} \cdot p_n^i - H\left(\Phi(P_n)\right) \right)$$

$$\leq H\left(\Phi(Q)\right) - \liminf_{n\to\infty} H\left(\Phi(P_n)\right)$$

$$= 0.$$

4. $\Rightarrow$ 1.

Let $C \subseteq M_+^1(U)$ be a set with $\Gamma(C, Q) > 0$. Then there exists a code improvement Δ such that $\Gamma(\Delta) \geq \varepsilon$. Therefore $\langle \Delta, P \rangle \geq \varepsilon$ for $P \in C$. Let f be the function $\frac{\lfloor k \cdot \Delta \rfloor}{k}$. Then $f \leq \Delta$

Assume that $\Phi(P_n) \xrightarrow{I} \Phi(Q)$ for all measurable maps $\Phi : (\Omega, \mathbb{A}) \to \mathbb{N}$ such that $\Phi(Q)$ is power bounded. Let $f : \mathbb{A} \to \mathbb{R}_+$ be a measurable function such that $\exp\left(t \cdot f(u)\right)$ is Q-integrable for some $t > 0$. Let Φ denote the function $\frac{\lfloor k \cdot f \rfloor}{k}$. Then

$$\int \exp\left(t \cdot f(u)\right) d\Phi Q u \geq \int \exp\left(t \cdot \frac{\lfloor k \cdot f(u) \rfloor}{k}\right) d\Phi(Q)u$$
$$\geq \int \left(\exp\left(\frac{\lfloor k \cdot f(u) \rfloor}{k}\right)\right)^t d\Phi(Q)u$$
$$\geq k^{-t} \int i^t \Phi(Q)\left(\frac{i}{k}\right)$$

Thus Φ is power bounded and

$$\left\langle \frac{\lfloor k \cdot f \rfloor}{k}, \Phi(P_n) \right\rangle \to \left\langle \frac{\lfloor k \cdot f \rfloor}{k}, \Phi(Q) \right\rangle$$

for all k. For all P the following inequality holds

$$\left\langle \frac{\lfloor k \cdot f \rfloor}{k}, \Phi(P) \right\rangle \leq \int f \, dP \leq \left\langle \frac{\lfloor k \cdot f \rfloor + 1}{k}, \Phi(P) \right\rangle = \left\langle \frac{\lfloor k \cdot f \rfloor}{k}, \Phi(P) \right\rangle + \frac{1}{k}$$

and therefore $\int f \, dP_n \to \int f \, dQ$. ■

Corollary 26. *Assume that Q is power dominated. Then $P_n \overset{I_s^*}{\to} Q$ if and only if $P_n \overset{I}{\to} Q$.*

Combining Corollary and Example 13 we see that the sequential weak information topology satisfies neither first nor second countability axiom.

To see that the sequential weak and the strong information topology are different we just have to remark that the I-closed set of atomic probability measures is dense in the set of all probability distributions because a probability distribution cannot be distinguished from a atomic distribution by any number of functions $\Delta_i, i = 1, \ldots, n$. An example of a sequence converging in τ_0 but not in I_s^* is given by the following construction. Let Q be the probability measure on $\mathbb{N}$ given by the point probabilities $Q(i) = 2^{-i}$. Then $\sum_{i=1}^{\infty} i \cdot Q(i) = 2 < \infty$ and therefore the function $f : i \curvearrowright \log i$ satisfies $\sum \exp\left(t \cdot f(i)\right) \cdot Q(i) < \infty$ for a $t > 0$. Let P_n be the probability measure given by

$$P_n(i) = \left(1 - \frac{1}{\log n}\right) Q(i) + \frac{1}{\log n} \delta_{i,n}.$$

Then P_n converges pointwize to Q and therefore also in τ_0. At the same time

$$\sum_{i=1}^{\infty} f(i) \cdot P_n(i) = \left(1 - \frac{1}{\log n}\right) \sum_{i=1}^{\infty} f(i) \cdot Q(i) + \frac{1}{\log n} \sum_{i=1}^{\infty} \log(i) \cdot \delta_{i,n}$$

$$= \left(1 - \frac{1}{\log n}\right) \sum_{i=1}^{\infty} f(i) \cdot Q(i) + 1$$

$$\to \sum_{i=1}^{\infty} f(i) \cdot Q(i) + 1 \text{ for } n \to \infty.$$

Let $A \subseteq M_+^1(U)$ be a set of probability distributions. Assume that A satisfy the following criterion.

Criterion 27. *For all $Q \in A$ there exists a finite number of sets $C_i \subseteq M_+^1(U)$ such that $\Gamma(C_i, Q) > 0$ and $\complement A \subseteq \bigcup C_i$.*

Then A is open in the sequential weak information topology. The set of sets satisfying Criterion 27 is obviously a topology which we will call the *weak information topology* and denote I^*.

Theorem 28. *The weak information topology is stronger than the topology τ_0, but weaker than the strong information topology.*

Proof. This follows since $\Gamma(C, Q) \leq D(C \parallel Q)$. ∎

The topologies τ_0 and I have the same connected components and therefore I^* has the same connected components as these topologies. Similarly, on a finite set the topologies τ_0 and I coincide and therefore they coincide with I^* on finite sets.

As immediate consequences of Theorem 28 we get that the map $(P, Q) \curvearrowright D(P \parallel Q)$ is lower semi continuous when $M_+^1(U)$ is equipped with the weak information topology. Especially the sets

$$\left\{P \in M_+^1(U) \mid D(P \parallel Q) \leq r\right\}$$

are I^*-closed.

We easily see that the weak information topology is compact if and only if U is a singleton.

Proposition 29. *For the weak information topology the following conditions are equivalent.*

1. *U is finite.*
2. *$M_+^1(U)$ is σ−compact.*
3. *$M_+^1(U)$ is locally compact.*

Proof. 1. ⇒ 2. Obvious.

2. ⇒ 3. Obvious.

¬3. ⇒ ¬1. Assume that U is not finite, and let Q be a probability measure with infinite support. Assume that ω is a compact neighborhood in weak information topology. Then ω contains a ball $B(Q,r) \subseteq \omega$, and the balls are closed so $B(Q,r)$ is compact. Now there exist a sequence $P_n \to Q$ in I such that P_n has finite support. Therefore $B(Q,r)$ contains a probability distribution P with finite support. The sequence $\left(1-\frac{1}{n}\right)P+\frac{1}{n}Q \to P$ in weak topology but not in τ_0 and therefore not in the weak information topology. Therefore the sequence $\left(1-\frac{1}{n}\right)P+\frac{1}{n}Q$ has no convergent subnet, and the contradiction is obtained. ■

Proposition 30. *The weak information topology is separable if and only if U is countable.*

Proof. If U is finite the weak information topology is obviously separable.

If U is countable then $M_+^1(U)$ is separable in the strong information topology and therefore also in the weak information topology.

If U is not countable, then $M_+^1(U)$ is not separable in τ and therefore not in the weak information topology. ■

The topology τ and the topology corresponding to total variation can be extended to the set of bounded distribution with sign. The set of bounded measures is a group and both τ and the topology corresponding to total variation are translation invariant in this group. None of the information topologies extends into translation invariant topologies on the set of bounded signed measures. We are interested in convergence theorems and are therefore primarily interested in the topology near a point $Q \in M_+^1(U)$. Instead of having a topology which is not translation invariant one could form a translation invariant topology by taking the local structure (neighborhoods) of I^* near Q and translate it to get neighborhoods of other points. Although not explicitly stated this is the technique used in [12].

5. The conditional limit theorem

The idea to use information theoretical methods to prove a strong version of the conditional limit theorem was given by I. Csiszár in [8]. He introduced the topology τ_0 in order to formulate the conditional limit theorem with a stronger topology than τ. We will follow these ideas and see to what extend the information topologies are useful in formulating a conditional limit theorem.

Let $(U, \mathbb{F}, Q)$ be a probability space with an *prior* distribution Q. Throughout this section we consider a fixed set $C \subseteq M_+^1(U)$ of consistent distributions. We will show that the principle of minimum information discrimination can be interpreted as an extension of the concept conditional probability. Let $\omega = (u_1, u_2, \ldots, u_n) \in U^n$ be a sample, i.e. a finite sequence of outcomes, then the empirical distribution is $\mathrm{Emp}_n(\omega) = \frac{1}{n}\sum_{i=1}^n \delta_{u_i}$, where δ_{u_i} is the Dirac measure in the point u_i. On a finite set the empirical distribution is also called the type of the sequence [9]. We wish to find the *posterior* probability distribution under the condition that the empirical distribution is consistent, i.e. the empirical distribution is element in C. Thus, the situation is that for some reason we cannot observe the empirical distribution directly, but using some indirect observation we can infer that the empirical distribution belong to the set C. An example from statistical mechanics is a temperature measurement which tells you the mean energy of the molecules but not the energy of any individual molecule.

Define

$$K_n = \{\omega \in \Omega \mid \mathrm{Emp}_n(\omega) \in C\}.$$

If K_n is $\mathbb{F}^n$-measurable and $Q^n(K_n) > 0$, it is possible to consider the conditional probability of Q^n with respect to the set K_n in U^n. Let $Q_n = Q^n(\cdot \mid K_n)_{(1)}$ denote the marginal distribution of $Q^n(\cdot \mid K_n)$. An alternative way to write Q_n is

$$E\big(\,\mathrm{Emp}_n(\omega) \mid K_n\big)$$

in the sense that

$$\langle f, Q_n\rangle = E\big(\langle f, \mathrm{Emp}_n(\omega)\rangle \mid K_n\big)$$

for all measurable functions f. This follows from the calculation

$$E\big(\operatorname{Emp}_n(\omega) \mid \operatorname{Emp}_n(\omega) \in C\big) = \frac{1}{n} \cdot \sum_{j=1}^{n} E\big(\delta_{u_j} \mid \operatorname{Emp}_n(\omega) \in C\big)$$
$$= E\big(\delta_{u_1} \mid \operatorname{Emp}_n(\omega) \in C\big)$$
$$= Q^n(\cdot \mid K_n)_{(1)}.$$

Theorem 31. *Let (U, Q) be a probability space, and let C be a set of probability measures on U.*

If $K_n \in \mathbb{F}^n$ and $Q^n(K_n) > 0$, define $Q_n = Q^n(\cdot \mid K_n)_{(1)}$. If Δ is a measurable function satisfying $\int \exp(\Delta)\, dQ \leq 1$ then

(3) $$\langle \Delta, Q_n \rangle \geq \Gamma(\Delta).$$

For any set $C \subseteq M_+^1(U)$ the following inequality holds

(4) $$\Gamma(C, Q) \leq -\frac{1}{n} \log P^*(K_n)$$

where $P^*\big(\operatorname{Emp}_n(\omega) \in C\big)$ denotes the outer measure of the set

$$\big\{\omega \in U^n \mid \operatorname{Emp}_n(\omega) \in C\big\}.$$

If Q is in equilibrium with C the following inequality holds

$$D(C \parallel Q) \leq -\frac{1}{n} \log P^*(K_n).$$

Proof. Assume that $K_n = \big\{\omega \in U^n \mid \operatorname{Emp}_n(\omega) \in C\big\}$ is $\mathbb{F}^n$-measurable and that $Q^n(K_n) > 0$.

$$\langle \Delta, Q_n \rangle = E\big(\big\langle \Delta, \operatorname{Emp}_n(\omega)\big\rangle \mid K_n\big)$$
$$\geq E\big(\Gamma(\Delta)\big)$$
$$= \Gamma(\Delta).$$

To get (4) first assume that K_n is $\mathbb{F}^n$-measurable. Then

$$D(Q_n \parallel Q) \geq \langle \Delta, Q_n \rangle \geq \Gamma(\Delta).$$

This implies that

$$\begin{aligned} P\big(\,\mathrm{Emp}_n(\omega) \in C\big) &= Q^n(K_n) \\ &= \exp\big(-D\big(Q^n(\cdot \mid K_n) \,\|\, Q^n\big)\big) \\ &\le \exp\big(-nD(Q_n \,\|\, Q)\big) \\ &\le \exp\big(-n\Gamma(\Delta)\big), \end{aligned}$$

where the sub-additivity of the information divergence is used to obtain the first inequality. This holds for all measurable functions Δ with $\int \exp(\Delta)\, dQ \le 1$ and therefore

$$P\big(\,\mathrm{Emp}_n(\omega) \in C\big) \le \exp\big(-n\Gamma(C, Q)\big).$$

Now assume that $K_n \notin \mathbb{F}^n$. Put

$$\tilde{K}_n = \left\{ \omega \in U^n \,\middle|\, \left(\frac{1}{n}\sum_{i=1}^{n} \Delta_i\right)(u_1, u_2, \ldots, u_n) \ge \Gamma(C, Q) \right\}.$$

Then $\tilde{K}_n$ is measurable because each of the functions Δ_i are measurable. Then

$$\begin{aligned} P^*\big(\,\mathrm{Emp}_n(\omega) \in C\big) &\le P\big(\,\mathrm{Emp}_n(\omega) \in \tilde{K}_n\big) \\ &\le \exp\big(-n\Gamma(C, Q)\big), \end{aligned}$$

which proves Inequality 4.

If Q is in equilibrium with C then

$$\Gamma(C, Q) = D(C \,\|\, Q),$$

and the last inequality holds. ■

The quantities in (4) are often equipped with the opposite sign turning the inequality into an upper bound. As the inequality is stated here

$$-\log P^*\big(\,\mathrm{Emp}_n(\omega) \in C\big)$$

has a natural interpretation as a mean code length. The following theorem is a law of large numbers related to the weak information topology.

Theorem 32. *Let $(U, \mathbb{F}, Q)$ be a probability space, and $(\Omega, \mathbb{G}, P)$ is the corresponding probability space of the sequences such that $\Omega = (U^{\mathbb{N}}, \mathbb{G})$ the σ-algebra generated by the cylinder sets, and P a probability measure such that $P_{|\mathbb{F}^n} = Q^n$. Let A be a I^*-neighborhood of the distribution Q. Then the empirical distribution $\mathrm{Emp}_n(\omega) \in A$ eventually P-almost surely. In particular the empirical distribution I^*-converges to Q in probability.*

Proof. For any set C with $\Gamma(C, Q) > 0$ and in equilibrium we have

$$\begin{aligned} P\big(\,\mathrm{Emp}_n(\omega) \in C \text{ for some } n \geq N\big) &\leq \sum_{n \geq N} P(K_n) \\ &\leq \sum_{n \geq N} \exp\big(-n\Gamma(C, Q)\big) \\ &\leq \frac{\exp\big(-N\Gamma(C, Q)\big)}{1 - \exp\big(-\Gamma(C, Q)\big)}. \end{aligned}$$

This shows that the probability that the empirical distribution lies in C for some $n \geq N$ converges to 0 for N going to ∞. ■

For any topology weaker than the weak information topology and satisfying first countability axiom one gets almost sure convergence. To see this let ω_m be a countable basis of the neighborhoods of Q is such a topology. Then the set B_m of sequences with $\mathrm{Emp}_n(\omega) \in \omega_m$ eventually has measure 1. Therefore also the set $\bigcap_{m=1}^{\infty} B_m$ has measure 1.

Even on a finite set Theorem 32 gives a stronger law of large numbers than the formulations normally found in textbooks. The difference is that Theorem 32 implies that the support of the empirical distribution is a subset of the support of Q.

The approach presented here can also be used to give an asymptotic upper bound on $-\frac{1}{n} \log P\big(\,\mathrm{Emp}_n(\omega) \in C\big)$, but only in the case where C is a "fat" set in the sense that C contains inner points. For the "thin" case see [19] and references in there.

Lemma 33. *Let $(U, \mathbb{F}, Q)$ be a probability space, and let C be a set of probability measures on U. Then*

$$\limsup_{n \to \infty} -\frac{1}{n} \log P_*\big(\,\mathrm{Emp}_n(\omega) \in C\big) \leq D(\mathrm{int}_{I^*} C \parallel Q) \tag{5}$$

where $P_*\big(\operatorname{Emp}_n(\omega) \in C\big)$ *denotes the inner measure of the set*

$$\{\omega \in \Omega \mid \operatorname{Emp}_n(\omega) \in C\}$$

with respect to the probability measure $P = Q^n$.

Remark 34. In this lemma C is *not* assumed to be convex.

Proof. Let R be an I^*-inner point in C.

First assume that $K_n = \{\omega \in U^n \mid \operatorname{Emp}_n(\omega) \in C\}$ is $\mathbb{F}^n$-measurable. Then

$$\begin{aligned} -\frac{1}{n}\log P(K_n) &= -\frac{1}{n}\log\big(Q^n(K_n)\big) \\ &= \frac{1}{n}D\big(Q^n(\cdot \mid K_n) \,\|\, Q^n\big) \\ &\leq \frac{1}{n}D\big(R^n(\cdot \mid K_n) \,\|\, Q^n\big) \end{aligned}$$

because the conditional probability measure $Q^n(\cdot \mid K_n)$ minimizes the divergence among measures concentrated on K_n.

We use the equality

$$\begin{aligned} D(R^n \,\|\, Q^n) = {} & R^n(K_n) \cdot D\big(R^n(\cdot \mid K_n) \,\|\, Q^n\big) \\ & + R^n(\complement K_n) \cdot D\big(R^n(\cdot \mid \complement K_n) \,\|\, Q^n\big) \\ & - H\big(R^n(K_n), R^n(\complement K_n)\big), \end{aligned}$$

which leads to the inequality

$$D\big(R^n(\cdot \mid K_n) \,\|\, Q^n\big) \leq \frac{D(R^n \,\|\, Q^n) + H\big(R^n(K_n), R^n(\complement K_n)\big)}{R^n(K_n)}.$$

Therefore

$$\begin{aligned} -\frac{1}{n}\log P(K_n) &\leq \frac{1}{n}D\big(R^n(\cdot \mid K_n) \,\|\, Q^n\big) \\ &\leq \frac{D(R^n \,\|\, Q^n) + H\big(R^n(K_n), R^n(\complement K_n)\big)}{n \cdot R^n(K_n)} \\ &= \frac{D(R \,\|\, Q)}{R^n(K_n)} + \frac{H\big(R^n(K_n), R^n(\complement K_n)\big)}{n \cdot R^n(K_n)} \\ &\to D(R \,\|\, Q) \quad \text{for} \quad n \to \infty, \end{aligned}$$

because $R^n(K_n) \to 1$ for $n \to \infty$ by law of large numbers.

If K_n is not measurable then find a subset A of the form

$$A = \{ S \in M_+^1(U) \mid \langle \Delta_i, S \rangle \le c_i,\ i = 1, \dots, k \}$$

where c_i, $i = 1, 2, \dots, k$ are positive constants and and Δ_i functions satisfying $\int \exp(\Delta_i)\, dQ \le 1$. Then $\{ \omega \in U^n \mid \mathrm{Emp}_n(\omega) \in A \}$ is $\mathbb{F}$-measurable and

$$P_*(K_n) \le P\big(\mathrm{Emp}_n(\omega) \in A \big)$$

and

$$\begin{aligned} \limsup_{n\to\infty} -\frac{1}{n} \log P_*(K_n) &\le \limsup_{n\to\infty} -\frac{1}{n} \log P_*\big(\mathrm{Emp}_n(\omega) \in A \big) \\ &\le D(R \parallel Q). \end{aligned}$$

This holds for all $R \in \mathrm{int}_{I^*} C$ and the inequality follows. ∎

The theorem is somewhat stronger than similar theorems in the literature as the weak information topology is stronger than the topology τ_0 used in the literature. By combining the lower and upper bounds one gets

$$\begin{aligned} \Gamma(C, Q) &\le -\frac{1}{n} \log P^*(K_n) \\ &\le \limsup_{n\to\infty} -\frac{1}{n} \log P_*\big(\mathrm{Emp}_n(\omega) \in C \big) \\ &\le D(\mathrm{int}_{I^*} C \parallel Q). \end{aligned}$$

Thus the lower bound refers to one side of the game and the upper bound refers to the other side of the game. The following theorem includes the well-known Conditional Limit Theorem.

Theorem 35. *Let $(U, \mathbb{F}, Q)$ be a probability space, and let C be a convex set in equilibrium with Q such that*

$$D(C \parallel Q) = D(\mathrm{int}_{I^*} C \parallel Q) < \infty.$$

If K_n is $\mathbb{F}^n$-measurable for all n then $Q^n(K_n) > 0$, eventually. Put $Q_n = Q^n(\cdot \mid K_n)_{(1)}$. Then $Q_n \xrightarrow{I} Q_{|C}$ for $n \to \infty$. If we condition on $\mathrm{Emp}_n(\omega) \in C$, the $\mathrm{Emp}_n(\omega) \xrightarrow{I^*} Q_{|C}$ in probability.

Finally

$$-\frac{1}{n}\log P(K_n) \to D(C \parallel Q) \text{ for } n \to \infty$$

in the sense that both inner and outer Q^n*-measure of* $\{\omega \in \Omega \mid \mathrm{Emp}_n(\omega) \in C\}$ *converges to* $D(C \parallel Q)$.

Proof. First remark that $-\frac{1}{n}\log P\big(\mathrm{Emp}_n(\omega) \in C\big) \le D(P \parallel Q) + \varepsilon$ eventually, which shows that $P\big(\mathrm{Emp}_n(\omega) \in C\big) > 0$ eventually. Using subadditivity of the divergence we get

$$\begin{aligned} D(Q_n \parallel Q) &\le \frac{1}{n} D\big(Q^n(\cdot \mid K_n) \parallel Q^n\big) \\ &= -\frac{1}{n}\log P(K_n), \end{aligned}$$

Combining Theorem 31 and Lemma 33 proves that

$$D(Q_n \parallel Q) \searrow D(C \parallel Q) \quad \text{for} \quad n \to \infty.$$

Using the inequality in Theorem 1 we get that $Q_n \xrightarrow{I} Q_{|C}$ for $n \to \infty$. ■

The condition that C is in equilibrium does not cover all cases where the conditional limit theorem holds, but on a countable space one can replace C with $C_0 = \{P \in C \mid D(P \parallel Q) < \infty\}$. Now C_0 is in equilibrium with Q. All empirical measures $\mathrm{Emp}_n(\omega) \in C$ also satisfy $\mathrm{Emp}_n(\omega) \in C_0$ because empirical measures are atomic. Therefore $D(C_0 \parallel Q) = D(C \parallel Q)$ and the sequence Q_n is the same when C is replaced by C_0.

If C is not in equilibrium then we obtain the following lower bound

$$D(\mathrm{cl}_\tau C \parallel Q) \le \liminf -\frac{1}{n} \cdot \log P(K_n).$$

The bound is obvious if C is a finite union of convex sets in equilibrium. In the general case we use that the information balls $B(Q,r)$ are τ-compact. This lower bound is very weak as we know that in general Q is element in the closure of $\complement B(Q,r)$ even in the strong information topology.

6. Discussion

In [8] I. Csiszár used a convexity condition called almost absolute convexity in order to get a conditional limit theorem. Let C be a set and Q a distribution. For a code improvement Δ with $\Gamma(\Delta) < \infty$ the convex set $\{P \in M_+^1(U) \mid \int \Delta \, dP \geq \Gamma(\Delta)\}$ is absolute convex. Then also the intersection of all such absolute convex sets is absolutely convex, and this intersection is in equilibrium if and only if the original set was in equilibrium. Therefore the concept of game theoretical equilibrium can "replace" Csiszár's convexity condition. A code improvement Δ defines a measure R with density $dR/dQ = \exp(\Delta)$. For a given measure R in general there may exist several code improvements satisfying the equation. Let Δ_1 and Δ_2 be two such code improvements. In general the sets $\{P \in M_+^1(U) \mid \int \Delta_i \, dP \geq \Gamma(\Delta_i)\}$, $i = 1, 2$ will be different. Therefore it is important to consider code improvements as functions and not just to identify the code improvement with the corresponding measure.

The sequence $Q_n \xrightarrow{I^*} Q$ if and only if for any given statistical test we are eventually not able to distinguish between Q and the elements in the sequence. Or more loosely: A sequence converges in the weak information topology if and only if the elements are asymptotically indistinguishable. The sequence $Q_n \xrightarrow{I} Q$ if and only if we are eventually not able to distinguish between Q and the elements in the sequence uniformly on all statistical test corresponding to a given level of the divergence. This is related to universal testing, where one uses a test which asymptotically can distinguish a distribution from any other distribution. Thus, a sequence converges in the strong information topology if and only if the elements are asymptotically universally indistinguishable.

In this paper two weak information topologies have been introduced. In general they are different, and there exists nets convergent in the weak information topology but not in the sequential weak information topology. It is not clear whether there exists sequences converging in I^* but not in I_s^*. If no such sequences exists one could simply define I_s^* as the sequential topology with the same convergent sequences as I^*.

As the minimum information game is described in this paper the generalized information projection of Q on a convex set C exists if and only if $D(C \parallel Q) < \infty$. In some cases an optimal code improvement may exist even in the case where $D(C \parallel Q) = \infty$, and it is of interest to study conditions for an optimal code improvement to exist.

If the set of outcomes is partitioned, then the strong information topologies can be considered on the partitions. For a set of partitions one can consider the initial topology corresponding the set of partitions. This will give a topology which is weaker than the strong topology. Topologies corresponding to partitions seems to play a an important role in a deeper understanding of the conditional limit theorem, and its relation to a "weak information projection". A detailed treatment of these questions is postponed to a future paper.

In [17] discontinuity of the entropy function was used to explain why Zipf's law is applicable in linguistics. In a continuous setting one should expect that the result that information divergence $P \curvearrowright D(P \parallel Q)$ is discontinuous at P_0 if and only if the random variable $\frac{dP_0}{dQ}$ is hyperbolic, could be used to explain the appearance of power laws and other heavy tailed distributions in certain cases.

Finally we remark that it is possible to extend many of the definitions and results on the information topologies to a non-commutative setting where distributions are replaced by density operators on a Hilbert space. In a non commutative setup the Bloch sphere will be disconnected in the (strong) information topology, and the components will be the interior and each of the points on the boundary.

Acknowledgement. The author thanks Jan Caesar, Flemming Topsøe and Imre Csiszár for useful discussions. Flemming Topsøe pointed my attention to general results on existence of topologies defined by sequences. After the first submission of this paper I had useful discussions with Richard Dudley where he pointed my attention to his paper [11]. Mikhail B. Malioutov has also given many comments which have improved this paper. The example of a set for which n closure operations are needed in order to get the closure was developed together with Franšitek Matúš during a visit on his institute. Finally I will thank an anonymous reviewer for many relevant comments and suggestions for improvements.

References

[1] S. Amari, Information geometry on hierarchy of probability distributions, *IEEE Trans. Inform. Theory,* **47** (2001), 1701–1711.

[2] A. R. Barron, Entropy and the Central Limit Theorem, *Annals Probab. Theory,* **14(1)** (1986), 336–342.

[3] A. R. Barron, Limits of information, Markov chains, and projections, in: *Proceedings 2000 International Symposium on Information Theory* (2000), p. 25.

[4] T. Cover and J. A. Thomas, *Elements of Information Theory* (1991), Wiley.

[5] I. Csiszár, Informationstheoretische Konvergenzbegriffe im Raum der Wahrscheinlichkeitsverteilungen, *A Magyar Tudományos Akadémia Matematikai Kutató Intézetének Közleményei,* **7** (1962), 137–158.

[6] I. Csiszár, Über topologische und metrische Eigenscaften der relativen Information der Ordnung alpha, in: *Transactions of the Third Prague Confernce on Information Theory, Statistical Decision Functions, Random Processes, 1962* (1964), pp. 63–73, Prague. Publishing House of the Czechoslovak Academy of Science.

[7] I. Csiszár, I-divergence geometry of probability distributions and minimization problems, *Ann. Probab.,* **3** (1975), 146–158.

[8] I. Csiszár, Sanov property, generalized I-projection and a conditional limit theorem, *Ann. Probab.,* **12** (1984), 768–793.

[9] I. Csiszár, The method of types, *IEEE Trans. Inform. Theory,* **44(6)** (Oct. 1998), 2505–2523.

[10] A. Dembo and O. Zeitouni, *Large Deviations Techniques and Applications,* Jones and Bartlett Publishers International (Boston, 1993).

[11] R. Dudley, Consistency of m-estimators and one-sided bracketing, in: *High Dimensional Probability,* E. E. et al., editor, Progress in Probability, Birkhäuser (Basel, 1998), pp. 33–58.

[12] P. Eichelsbacher and U. Schmock, Large deviations for products of empirical measures of dependent sequences, *Markov Processes and Related Fields,* **7(3)** (2001), 435–468.

[13] R. Engelking, *General Topology,* volume 6 of *Sigma Series in Pure Mathematics,* Heldermann (Berlin, 1989), revised and completed edition.

[14] J. Fritz, An information-theoretical proof of limit theorems for reversible Markov processes, in: *Trans. Sixth Prague Conf. on Inform. Theory, Statist. Decision Functions, Random Processes,* Prague, 1973. Czech. Acad. Science, Academia Publ. (Prague, Sept. 1971).

[15] G. H. Hardy and M. Riesz, *The general Theory of Dirichlet's series,* Cambridge Univ. Press (London, 1915).

[16] P. Harremoës, Binomial and Poisson distributions as maximum entropy distributions, *IEEE Trans. Inform. Theory,* **47(5)** (July 2001), 2039–2041.

[17] P. Harremoës and F. Topsøe, Maximum entropy fundamentals, *Entropy,* **3(3)** (Sept. 2001), 191–226. `http://www.unibas.ch/mdpi/entropy/` [ONLINE].

[18] E. T. Jaynes, Information theory and statistical mechanics, I and II, *Physical Reviews,* **106** (1957), 620–630 and **108** (1957), 171–190.

[19] O. Johnson, Entropy and a generalisation of Poincare's observation, *Mathematical Proceeding of the Cambridge Philosophical Society,* **135(2)** (2003), 375–384.

[20] D. G. Kendall, Information theory and the limit theorem for Markov chains and processes with a countable infinity of states, *Ann. Inst. Stat. Math.*, **15** (1964), 137–143.

[21] J. Kisynski, Convergence du typè l, *Colloq. Math.*, **7** (1960), 205–211.

[22] S. Kullback, *Information Theory and Statistics,* Wiley (New York, 1959).

[23] Y. V. Linnik, An information-theoretic proof of the central limit theorem with Lindeberg condition, *Theory Probab. Appl.*, **4** (1959), 288–299.

[24] A. W. Marshall and I. Olkin, *Inequalities: Theory of Majorization and its Applications.* Academic Press (New York, 1979).

[25] P. Ney, Dominating points and the asymptotics of large deviations for random walks on R^n, *Ann. Probab.*, **11(1)** (1983), 158–167.

[26] A. Rényi, On measures of entropy and information, in: *Proc. 4th Berkeley Symp. Math. Statist. and Prob.*, volume **1**, Univ. Calif. Press. (Berkeley, 1961), pp. 547–561.

[27] S. Takano, Convergence of entropy in the central limit theorem, *Yokohama Mathematical Journal,* **35** (1987), 143–148.

[28] F. Topsøe, Information theory at the service of science, in this volume.

[29] F. Topsøe, Information theoretical optimization techniques, *Kybernetika,* **15(1)** (1979), 8–27.

Peter Harremoës

Centrum voor Wiskunde en Informatica
Kruislaan 4/3
NL-1098 SJ Amsterdam
P.O.Box 94079
The Nederlands

harremoes@ieee.org

BOLYAI SOCIETY
MATHEMATICAL STUDIES, 16

Entropy, Search, Complexity, pp. 151–158.

Reinforced Random Walk

MICHAEL KEANE

One of the distinguishing properties of the present scientific method is reproducibility. In one of its guises, probability theory is based on statistical reproduction, near certainty being obtained of truth of statements by averaging over long term to remove randomness occurring in individual experiments.When one assumes, as is often the case, that events farther and farther in the past have less and less influence on the present, the probabilistic paradigm is currently well understood and is successful in many scientific and technological applications. Recently, however, we have come to realize that precisely in these applications important stocahstic processes occur whose present outcomes are significantly influenced by events in the remote past. This behaviour is not at all well understood and some of the simplest questions remain today irritatingly beyond reach. A salient example occurs in the theory of random walks, where there is a dichotomy between recurrent and transient behaviour. After explaining this classical dichotomy, we present a very simple example with infinite memory which is neither known to be transient nor recurrent. Then, using a reinforcement mechanism due to Pólya, we explain the nature of a particular infinite memory process in terms of spontaneous emergence of opinions. Finally we would like to discuss briefly some of our recent results towards understanding the recurrence-transience dichotomy for reinforced random walks.

An open problem

First consider the graph whose vertices are the points of $\mathbb{Z}^d$ and whose edges are the subsets $\{z, z'\}$ of $\mathbb{Z}^d$ with $|z-z'| = 1$, $|.|$ denoting Euclidean distance. Here d is a fixed positive integer. Let S_n, $n \geq 0$, be the position of a simple random walker on this graph:

- $S_0 = 0$ (the origin of $\mathbb{Z}^d$)
- $|S_{n+1} - S_n| = 1$ for each $n \geq 0$

- the *jumps* $X_{n+1} = S_{n+1} - S_n, n \geq 0$, are independent identically distributed random variables whose common distribution is uniform over the $2d$ possible elements in $\mathbb{Z}^d$ of length one.

In general, a discrete time and space stochastic process S_n, $n \geq 0$, is called *recurrent* if

$$\mathbb{P}(n \geq 1 : S_n = S_0) = 1),$$

and *transient* if this probability is less than one. We recall at this point a remarkable result due to Pólya:

- If $d = 1$ or $d = 2$, simple random walk on $\mathbb{Z}^d$ is recurrent.
- If $d \geq 3$, simple random walk on $\mathbb{Z}^d$ is transient.

As simple random walk is a Markov process, this yields a strong dichotomy:

- If $d = 1$ or $d = 2$, each point of $\mathbb{Z}^d$ is visited an infinite number of times by S_n, with probability one.
- If $d \geq 3$, each point of $\mathbb{Z}^d$ is visited at most finitely often by S_n, with probability one; that is, $\lim_{n\to\infty} S_n = \infty$ with probability one.

Our open problem consists of a seemingly slight modification of simple random walk on $\mathbb{Z}^2$. We denote the modified stochastic process by $S'_n, n \geq 0$. We still require

- $S_0 = 0$ (the origin of $\mathbb{Z}^d$)
- $|S_{n+1} - S_n| = 1$ for each $n \geq 0$
- the *jumps* $X_{n+1} = S_{n+1} - S_n$, $n \geq 0$, are distributed over the four possible elements of $\mathbb{Z}^2$ of length one.

However, these jumps X'_{n+1} will no longer be uniformly distributed; their distribution will depend on the history $\{S'_0, \ldots, S'_n\}$ of our process. The idea is to make the probability of traversing an edge at time $n+1$ larger than $1/4$ if the edge was previously visited, and less than $1/4$ if the edge was not previously traversed. A simple mechanism for this is to introduce a parameter $\theta > 1$, fixed, and to require that this probability is proportional to θ for edges previously traversed and proportional to 1 for edges not previously traversed.

For example, if $\theta = 2$ and if we arrive at a point $z \in \mathbb{Z}^2$ at time n, having previously traversed two of the four edges containing z (in either direction), then the probability of leaving z at time $n+1$ via one of these two edges is

$$\frac{\theta}{2+2\theta} = \frac{1}{3},$$

whereas the probability of leaving z via one of the two virgin edges is

$$\frac{1}{2+2\theta} = \frac{1}{6};$$

note that θ is simply the factor by which the probability of taking an edge is increased if it has previously been taken. Note also that if $\theta = 1$ we have the original simple random walk of Pólya in $\mathbb{Z}^2$, which is recurrent.

Open Problem. Prove that for some $\theta > 1$, the reinforced random walk S_n, $n \geq 0$, is recurrent.

We believe strongly that this is true for *any* $\theta > 1$, and in fact, it is easy to show that it is true in $d = 1$ for any $\theta > 0$. (If $\theta < 1$ then, strictly speaking, we should not use the word "reinforced", but this is a minor point.) However, the process S'_n is in no way similar to a Markov process, having memory which does not attenuate as time progresses, and up to now this problem remains beyond reach using currently known techniques.

Spontaneous emergence of opinions

In this section we attempt to give the reader a feeling for the naturality of questions concerning reinforcement. We feel this to be necessary for the following philosophical reason. The Markovian paradigm has been an extremely successful driving force for the development of probability theory over the last century, arguably the most important for practical applications. The main reason for this is that a large number of physical and technological type stochastic processes do have an asymptotic "forgetfulness" property – what happened in the past has less and less effect as time progresses, or, what happened very far away does not significantly influence the happenings here. This is a so-called locality principle which, if we exclude quantum mechanics, seems to be a law of nature; even in quantum probability there has been a large discussion and much disagreement concerning the possible validity of this type of locality. (As it now seems, the question has been decided through Bell's work and the Aspect experiments, but the discussion still continues.) Thus it seems to be useful to point to naturally arising processes which in no sense obey this principle. Here, I have chosen to illustrate the point with a story concerning spontaneous emergence of opinions; there are a number of parallel ideas in the realm of (universal) coding theory, and in this audience there are certainly people more capable than I of illustrating this phenomenon. What follows is not new mathematics, being based upon classical ideas again due to Pólya, developed around the 1920's – the important point is the interpretation in terms of reinforcement and the non-Markovian nature, leading to surprising behavior.

Let me begin by telling the story. Some 25 years ago I moved to The Netherlands and bought a house in Scheveningen, a bathing resort on the cost which is part of the town of The Hague, seat of government and royal residence. We knew at that point nothing about the surroundings. Our house was very close to the beach and also to the center of night life, so in the evening we had two natural alternatives for amusement:

- a visit to a nice bar B
- a stroll on the beautiful beach b

Thus we consider a graph with one vertex (our house) and two loops B (bar) and b (beach) from the vertex to itself. Our original *opinion* is denoted by a function giving *weights* to each of the loops – we start with each loop having weight one.

Each evening we search for amusement – for simplicity let us assume (but see the next section for more generality) that our search is restricted to B and b – and we choose one of these with a probability proportional to its current weight. Thus the first evening we visit the bar with probability 1/2 and the beach with probability 1/2. If, for example, we visit the bar and have a drink, we enjoy it very much, and decide to increase the weight of B by adding one to it – this would result in weight two for bar and weight one for beach. If the beach is visited, its weight will be increased by one.

After $n-2$ nights of entertainment, suppose that we have chosen $k-1$ times bar and $l-1$ times beach, with $k, l \geq 1$ and $k+l=n$. Then at this time, the bar weight is k and the beach weight is l, and the probability

$$\alpha_n = \frac{k}{n}$$

of choosing bar the next time is our *alcohol preference*, whereas the probability

$$\beta_n = 1 - \alpha_n = \frac{l}{n}$$

is our *nature preference*, or probability of choosing beach the next time. Thus our *opinion* at time n (after $n-2$ choices) is represented by the random pair

$$(\alpha_n, \beta_n).$$

It is a (not so simple but) interesting result of Pólya which tells us:

Our opinion becomes more and more stable as time passes.

or, in mathematical terms,

$$\alpha = \lim_{n\to\infty} \alpha_n$$

$$\beta = \lim_{n\to\infty} \beta_n$$

exist with probability one.

Nowadays we understand the reason behind this result to be contained in the *martingale convergence theorem*. That is, a simple calculation yields:

$$\mathbb{E}\left(\alpha_{n+1} \,\middle|\, \alpha_n = \frac{k}{n}\right) = \frac{k}{n} \cdot \frac{k+1}{n+1} + \frac{l}{n} \cdot \frac{k}{n+1} = \frac{k(n+1)}{n(n+1)} = \frac{k}{n} = \alpha_n$$

so that α_n is indeed a positive martingale, and the theory says that all positive martingales converge. (Pólya's proof was different, and perhaps simpler; it is not such an easy matter to prove martingale convergence theorems.)

Next comes a minor surprise: Suppose that not only I have bought a house in Scheveningen, but also a number of others, who search in the same manner for entertainment. Each of these householders develops a more and more stable opinion, but *these opinions differ in the limit.* For instance, perhaps in the limit

- α(Keane) $= 0.9$
- α(v.d.Toorn) $= 0.2$
- α(Pronk) $= 0.6$
- α(Hermina) $= 0.7$

- $\alpha(\text{Lootsma}) = 0.3$
- $\alpha(\text{Trahtenbroit}) = 0.8$

and so on. The second part of Pólya's result tells us the *distribution* of the limit opinion α:

The random limit α is uniformly distributed over the unit interval

Thus, although the alcohol preference of a given individual is random, we can predict the preferences of the population as a whole using its distribution, which is uniform! Some are alcoholic, some are nature lovers, and all types of mixtures occur equally often; each person is convinced of his or her own opinion.

This behavior is one of the salient characteristics of reinforced random walk. It is in fact very surprising that such a calculation can be accomplished, and in our open problem of the first section we know of no way to do this type of calculation. The reason that it is possible in this case is usually called *(partial) exchangeability.* This has been intensively studied by Diaconis and Friedman, and later for infinite graphs on trees by Pemantle. In our example things are very simple, which we now explain.

It is best to take a sample event. Suppose that for the first eight visits, three were visits to the beach and five were visits to the bar, in the following order:

B B b B b b B B

(B = bar, b = beach). Let us now calculate the probability of this event; after a bit of thought we see that it is

$$\frac{1}{2} \cdot \frac{2}{3} \cdot \frac{1}{4} \cdot \frac{3}{5} \cdot \frac{2}{6} \cdot \frac{3}{7} \cdot \frac{4}{8} \cdot \frac{5}{9} = \frac{3!5!}{9!}$$

(The reader should now verify this calculation step by step.

At this point, $n = 10$, $k = 6$, $l = 4$, and $\alpha_{10} = \frac{6}{10}$. The important point to notice is that if we have any other sequence of eight visits with three b's and five B's, then this probability remains the same. More generally, if n, kl, and l are given, then $\alpha_n = \frac{k}{n}$ and each sequence of $k-1$ B's and $l-1$ b's has probability

$$\frac{(k-1)!(l-1)!}{(n-1)!};$$

as there are

$$\frac{(n-2)!}{(k-1)!(l-1)!}$$

such sequences, we see that

$$\mathbb{P}\left(\alpha_n = \frac{k}{n}\right)$$

is the product of these two numbers, which is simply $\frac{1}{n-1}$. This is valid for each k in the range $1 \leq k \leq n-1$, which shows clearly that the limit distribution is uniform.

This concludes our philosophical section.

The current state of affairs

In this section I describe informally what we have done in the past few years concerning reinforcement. First of all, the result of Pólya for the simple case above has been extended to random walk with reinforcement on a finite graph with arbitrary initial weights, according to a suggestion of Diaconis and Coppersmith. This, together with related references, can be found in [1]. Sellke and Vervoort have studied intensively once reinforced random walks (as in the first section) on bi–infinite strips of finite widths, which are called ladders. The most recent results can all be found in the thesis of Vervoort [3]. It is still unknown whether for any value of the reinforcement parameter θ, once reinforced random walk is recurrent or not on ladders of width larger than two. The case of width two was settled a number of years ago by an interesting calculation due to T. Sellke (unpublished). In [2], we treat multiple reinforcement in essentially one-dimensional cases (tubes) when the weights are put on directed edges. Curiously enough, this is simpler and leads to relations with random walk in random environment. It is still unknown whether random walk with multiple reinforcement is recurrent or transient in two or higher dimensions, or whether the behavior depends on the amount of reinforcement. We do not discuss here the case in which the underlying graph is a tree, for which there are many interesting and nontrivial results.

References

[1] M. S. Keane and S. W. W. Rolles, Edge-reinforced random walks on finite graphs, in: *Infinite Dimensional Stochastic Analysis* (eds. Ph. Clément et al.), KNAW Verhandelingen, Afdeling Natuurkunde, Eerste Reeks, Deel **52**, (2000), pp. 217–234.

[2] M. S. Keane and S. W. W. Rolles, Tubular recurrence, *Acta Mathematica Hungarica,* **97** (2002), 207–221.

[3] M. Vervoort, *Games, walks and grammars: some problems I've worked on,* Dissertation, University of Amsterdam (2000).

Michael Keane

Department of Mathematics
Wesleyan University
Middletown, Connecticut 06459
U.S.A.

mkeane@wesleyan.edu

BOLYAI SOCIETY
MATHEMATICAL STUDIES, 16

Entropy, Search, Complexity, pp. 159–178.

QUANTUM SOURCE CODING AND DATA COMPRESSION*

DÉNES PETZ†

This lecture is intended to be an easily accessible first introduction to quantum information theory. The field is large and it is not completely covered even by the recent monograph [15]. Therefore the simple topic of data compression is selected to present some ideas of the theory. Classical information theory is not a prerequisite, we start with the basics of Shannon theory to give a feeling for Shannon entropy and for the informational divergence or relative entropy. The aim is to present Schumacher's compression theorem and to demonstrate that the von Neumann entropy, introduced in the 1920's by thermodynamical considerations, is a measure of quantum information exactly in the way as the Shannon entropy is that for classical information. Our discussion makes clear that the compression theorem depends heavily on the existence of the high-probability subspace.

At the end of the lecture quantum sources with memory and some related questions are briefly discussed. This part could be skipped by new-comers in the field.

1. CLASSICAL SOURCE CODING

Let X be a random variable with a finite range $\mathcal{X}$. A *source code* C for X is a mapping from $\mathcal{X}$ to the set of finite length strings of symbols of a D-ary alphabet which is assumed to be the set $\{0, 1, 2, \ldots, D-1\}$. Let $C(x)$ denote the codeword corresponding to x and let $\ell(x)$ denote the length of

---

*This text is a good written approximation of the first talk of a series given at the *Volterra-CIRM International School on Quantum information and quantum computing* in Trento, July, 2001.

†Partially supported by OTKA T032662 and T032374.

$C(x)$. If $p(x)$ is the probability of $x \in \mathcal{X}$, then the *expected length* of a source code C is given by

$$L(C) := \sum_x p(x)\,\ell(x).$$

Since the transmission of lengthy codewords could be costly, the aim of source coding is to make the expected code-length as small as possible. It is obvious that to meet this requirement the most frequent outcome of X must have the shortest codeword. For example in the Morse code the letter e (which is the most frequent one both in the English and Hungarian language) is represented by a single dot. (The *Morse code* uses an alphabet of four symbols: a dot, a dash, a letter space and a word space.) The extension of a code C from the finite length strings of $\mathcal{X}$ is defined by

$$C^*(x_1x_2\dots x_n) = C(x_1)C(x_2)\dots C(x_n),$$

where the right hand side is the concatenation of the corresponding codewords.

A code C is *uniquely decodable* if $C^*(x_1x_2\dots x_n) = C^*(x'_1x'_2\dots x'_m)$ implies that $x_1x_2\dots x_n = x'_1x'_2\dots x'_m$, that is $n = m$ and $x_i = x'_i$ for all $1 \le i \le n$. A code is called *prefix code* if no codeword is a prefix of any other. In case of a prefix code the end of a codeword is immediately recognised and hence such a code is uniquely decodable. For example, if 0, 10, 110 and 111 are the binary codewords (of a prefix code), then the binary string 1011001101110 is easily decomposed into 6 codewords: 10,110,0,110,111,0.

Theorem 1 (Kraft–MacMillan). *The codeword lengths $\ell(x)$ of a uniquely decodable code over an alphabet of size D satisfy the inequality*

$$\sum_x D^{-\ell(x)} \le 1.$$

Conversely, given a set of codeword lengths that satisfy this inequality, there exists a prefix code with these codewords lengths.

The proof is available in several standard books, for example [4]. It follows from the theorem that a uniquely decodable code could be always replaced by a prefix code which has the same codeword lengths.

Let $\lceil t \rceil$ denote the smallest integer $\ge t \in \mathbb{R}$. The codeword lengths $\ell(x) := \lceil -\log_D p(x) \rceil$ satisfy the Kraft inequality

$$\sum_x D^{-\ell(x)} \le \sum_x p(x) = 1.$$

According to the theorem there exists a prefix code with this codeword length. (Such a code is called *Shannon code.*) Since $-\log_D p(x) \le \ell(x) \le -\log_D p(x) + 1$, we have

$$-\sum_x p(x) \log_D p(x) \le L(C) \le 1 - \sum_x p(x) \log_D p(x).$$

for the expected code-length $L(C)$. For the rest we assume that $D = 2$. Then the bounds are given in terms of the *Shannon entropy* $H\big(p(x)\big) := -\sum_x p(x) \log p(x)$ as

$$H\big(p(x)\big) \le L(C) \le H\big(p(x)\big) + 1.$$

According to the next theorem the Shannon code is close to optimal.

Theorem 2. *The expected code-length of any prefix code is greater than or equal to the Shannon entropy of the source.*

Proof. We want to show $L - H\big(p(x)\big) \ge 0$ and estimate as follows

$$\begin{aligned} L - H(p) &= \sum_x p(x)\ell(x) + \sum_x p(x) \log p(x) \\ &= -\sum_x p(x) \log 2^{-\ell(x)} + \sum_x p(x) \log p(x) \\ &= \sum_x p(x) \log \frac{p(x)}{r(x)} - \log c, \end{aligned}$$

where $r(x) = c^{-1} 2^{-\ell(x)}$ and $c = \sum_x 2^{-\ell(x)}$. The *relative entropy* of two probability distributions is defined as

$$D(p\|r) := \sum_x p(x)\big(\log p(x) - \log r(x)\big)$$

and this quantity is known to be positive and 0 if and only if $p = q$. In terms of the relative entropy we have

$$L - H(p) = D(p\|r) + \log \frac{1}{c}.$$

Since $D(p\|r) \ge 0$ and $c \le 1$ from the Kraft–McMillan inequality, this shows $L - H(p) \ge 0$. ■

The Shannon code is close to optimal only if we know correctly the distribution of the source X. Assume that it is not the case and we associate to x the codeword length $\lceil -\log q(x)\rceil$, where q is another probability distribution on $\mathcal{X}$, possibly different from the true distribution p. One can compute that in this case

$$H(p) + D(p\|q) \le L(C) \le H(p) + D(p\|q) + 1. \tag{1}$$

For the use of the wrong distribution the relative entropy is the penalty in the expected length.

The optimal coding is provided by a procedure due to Huffman. The *Huffman code* is not easy to describe, therefore we show another coding due to Fano. The *Fano code* is nearly optimal, it satisfies the inequality

$$L(C) \le H(p) + 2.$$

In the Fano coding we order the probabilities $p(x)$ decreasingly as $p_1 \ge p_2 \ge p_3 \ge \ldots \ge p_m$. We choose k such that

$$\left|\sum_{i=1}^{k} p_i - \sum_{i=k+1}^{m} p_i\right|$$

is minimal. The division of the probabilities into the two classes divides the source symbols into two classes. A sign 0 for the first bit for the lower class and 1 for the first bit of the upper class. The two classes have nearly equal probabilities. Then we repeat the procedure for each of the two classes to determine the further bits of the code strings. This is Fano's scheme.

Up to now we have dealt with uniquely decodable codes. If the transmission of lengthy codewords is expensive, we might give up the exact decodability provided that the probability of mistake is small and long codewords can be avoided. This is a different approach to coding and decoding. Assume that the source emits the symbols $X_1, X_2, X_3, \ldots, X_n$ (independently and according to the same distribution p, typical for the source). We fix a coding procedure and all the emitted symbols are coded by this procedure which could be the Fano code, for example. Let $L_1, L_2, \ldots, L_n$ be the code-length of $X_1, X_2, \ldots X_n$, respectively. Both $X_1, X_2, \ldots, X_n$ and $L_1, L_2, \ldots, L_n$ are identically distributed independent random variables, the expectation of L_i is $L(C)$. The law of large numbers tells us that the probability of the event

$$L_1 + L_2 + \cdots + L_n \ge L(C) + \varepsilon \tag{2}$$

goes to 0 as $n \to \infty$. When $x_1, x_2, \ldots, x_n$ is a string of source symbols such that the corresponding code string is shorter than $n\bigl(L(C)+\varepsilon\bigr)$, then we code the string $x_1 x_2 \ldots x_n$ perfectly, otherwise we use always the same code string. If the latter case happens to occur, then we cannot recover the emitted symbol string from the code string. However, the probability of this error is exactly the probability of the event (2) which tends to 0. What did we win in this way? The number of source strings is $|\mathcal{X}|^n$ and the number of binary strings used in the coding is $2^{n(L(C)+\varepsilon)}$. When $L(C) < \log|\mathcal{X}|$, then

$$2^{n(L(C)+\varepsilon)} \ll |\mathcal{X}|^n.$$

Hence the cardinality of our code book is much smaller than the cardinality of the source strings if a small probability of error is allowed. We also say that that the data set $\mathcal{X}^n$ is compressed to a set of binary strings of length $n\bigl(L(C)+\varepsilon\bigr)$. What we have is an example of data compression. Efficient data compression is the same as source coding by short binary code strings. Since we need $n\bigl(L(C)+\varepsilon\bigr)$ binary digit for a source string of length n, $L(C)+\varepsilon$ is called *code rate.* (It is the number of binary digits needed for a single source symbol, in the average.) Using the Shannon code, we can achieve a code rate $H(p)+\varepsilon$. However, if we mistake the distribution of the source and assume q instead of p, then the rate is higher, it is about $H(p)+D(p\|q)+\varepsilon$. Hence the above method is very sensitive for the distribution of the source. To avoid this and to achieve slightly better code rate *block coding* can be used. Shortly speaking block coding means that the source string is not coded by letter by letter but the whole string gets a code string.

A *block code* $(2^{nR}, n)$ for a source $X_1, X_2, \ldots$ is given by two (sequences of) mappings:

$$f_n : \mathcal{X}^n \to \{1, 2, \ldots, 2^{nR_n}\}, \qquad \phi_n : \{1, 2, \ldots, 2^{nR_n}\} \to \mathcal{X}n.$$

Here f_n is the *encoder,* ϕ_n is the *decoder* and $R := \lim R_n$ is called the *rate of the code.* The *probability of error* of the code is

$$P_e^{(n)} := \text{Prob}\,\bigl(\phi_n \cdot f_n(X_1, \ldots, X_n) \neq (X_1, \ldots X_n)\bigr).$$

Shannon's source coding theorem is the following.

Theorem 3. *Let H be the entropy of the source and $R > H$. There exists a sequence of $(2^{nR_n}, n)$ block codes with error probability $P_e^{(n)}$ such that $P_e^{(n)} \to 0$ and $R_n \to R$.*

More precisely, this is only the positive part of Shannon's theorem telling that any rate $\geq H + \varepsilon$ is achievable under an arbitrary small bound on the probability of error. (The negative part tells that rates $< H$ are not achievable under the same constraint.)

Before we enter the proof we give an outline of the method of types. Let $\mathbf{x} \in \mathcal{X}^n$. The *type* of $\mathbf{x} = (x_1, x_2, \dots, x_n) \in \mathcal{X}^n$ is a probability mass function on $\mathcal{X}$. The mass of $x \in \mathcal{X}$ is the relative frequency of x in the sequence $(x_1, x_2, \dots, x_n)$:

$$P_{\mathbf{x}}(x) := \frac{1}{n} \#\{1 \leq i \leq n \, : \, x_i = x\}.$$

Let $\mathcal{P}_n$ denote the set of all types and for $P \in \mathcal{P}_n$ the *type class* of P is the set of all sequences of type P:

$$T(P) := \{\mathbf{x} \in \mathcal{X}^n \, : \, P_{\mathbf{x}} = P\}.$$

Since the frequency of any $x \in \mathcal{X}$ in a sequence $\mathbf{x} = (x_1, x_2, \dots, x_n)$ is at most n, we obviously have

$$\#(\mathcal{P}_n) \leq (n+1)^{\#(\mathcal{X})}.$$

The cardinality of a type class $T(P)$ is a multinomial coefficient but the following exponential bounds are useful:

$$\frac{1}{(n+1)^{\#(\mathcal{X})}} 2^{nH(P)} \leq \#\big(T(P)\big) \leq 2^{nH(P)}.$$

(A proof could be based on Stirling's formula on factorial functions, see [4] p. 282 for other proofs.)

Assume that a probability measure Q is given on $\mathcal{X}$ and let Q^n be the product measure on $\mathcal{X}^n$. The probability of a sequence $\mathbf{x} \in \mathcal{X}^n$ depends only on the type $P_{\mathbf{x}}$ of $\mathbf{x}$. A straight calculation gives that

$$Q^n\big(\{\mathbf{x}\}\big) = \prod_x Q(x)^{nP_{\mathbf{x}}(x)} = 2^{-nH(P_{\mathbf{x}}) + nD(P_{\mathbf{x}} \| Q)}.$$

The probability of a type class has exponential bounds:

$$\frac{1}{(n+1)^{\#(\mathcal{X})}} 2^{-nD(P\|Q)} \leq Q^n\big(T(P)\big) \leq 2^{-nD(P\|Q)}$$

for $P \in \mathcal{P}_n$.

Proof of Theorem 3. Let $\{Q(x) : x \in \mathcal{X}\}$ be the probability distribution of the given source and assume that $R > H(Q)$. Following the idea of Csiszár and Körner [5], we set

$$R_n := R - \#(\mathcal{X})\frac{\log(n+1)}{n}$$

and

$$A_n := \{\mathbf{x} \in \mathcal{X}^n : H(P_\mathbf{x}) \leq R_n\}.$$

Then

$$\#(A_n) = \sum \#(T(P)) \leq \sum 2^{nH(P)} \leq \sum 2^{nR_n}$$
$$\leq (n+1)^{\#(\mathcal{X})} 2^{nR_n} = 2^{nR},$$

where all summations are over the set $\{P \in \mathcal{P}_n : H(P) \leq R_n\}$. We can easily define an encoding and a decoding such that elements of A are encoded correctly and the other sequences give an error. (We just use elements of A as codewords). Then the probability of error is

$$P_e^{(n)} = 1 - \text{Prob}\,(A_n) = \sum Q^n(T(P)),$$

where the summation is over all $P \in \mathcal{P}_n$ such that $H(P) > R_n$. Estimating the sum by the largest term we obtain

$$P_e^{(n)} \leq (n+1)^{\#(\mathcal{X})} 2^{-n \min D(P\|Q)},$$

where min is over all $P \in \mathcal{P}_n$ such that $H(P) > R_n$. When n is large enough then $R_n > H(Q)$ and $Q \notin \{P \in \mathcal{P}_n : H(P) \leq R_n\}$. The minimum in the exponent is strictly positive and we can conclude that the probability of error converges to 0 exponentially fast as $n \to \infty$.

The interesting feature of the block code constructed in the proof of the theorem the fact that the distribution Q of the source does not appear, only its entropy $H(Q)$ should be known to construct the *universal encoding scheme.*

2. Quantum mechanical sources

A pure state of a quantum mechanical system is given by a unit vector of a Hilbert space. Assume that a quantum mechanical source emits the pure states $|\varphi_i\rangle$ with probability p_i $(1 \le i \le m)$. The source is specified by $\big(p_i, |\psi_i\rangle\big)_{i=1}^{m}$ which is called an *ensemble of* (pure) *quantum states.* Pure states of a quantum system are infinitely many and possess a fine topological structure. If after encoding and decoding we arrive at a state $|\psi_i'\rangle$ instead of $|\psi_i\rangle$, our error could be small when the vectors $|\psi_i\rangle$ and $|\psi_i'\rangle$ are close enough. Hence the problem of source coding in the quantum setting is rather different from the theory of source coding for finite classical sources and it is conceptually closer to the rate distortion theory initiated by Shannon as well.

How close are two quantum states? There are many possible answers to this question. Restricting ourselves to pure states, we have to consider two unit vectors. $|\varphi\rangle$ and $|\psi\rangle$. Quantum mechanics has used the concept of transition probability $\big|\langle\varphi \mid \varphi\rangle\big|^2$ for a long time (see cite, for example). This quantity is phase invariant, it lies between 0 and 1. It equals to 1 if and only if the two states coincide that is, $|\varphi\rangle$ equals to $|\psi\rangle$ up to a phase.

We call the square root of the transition probability *fidelity:*

$$F\big(|\varphi\rangle, |\psi\rangle\big) := \big|\langle\varphi \mid \psi\rangle\big|.$$

Shannon used a nonnegative distortion measure, and we may regard $1 - F\big(|\varphi\rangle, |\psi\rangle\big)$ as a distortion function on quantum states.

Under quantum operation a pure state could be transformed into a mixed state, hence we need extension of the fidelity:

$$F\big(|\varphi\rangle\langle\varphi|, D\big) = \sqrt{\langle\varphi \mid D \mid \varphi\rangle}.$$

(Some properties of fidelity are summarised in the Appendix.)

The n-shot source is $\big(p_I, |\psi_I\rangle\big)$ on the n-fold tensor product $\mathcal{H}^n = \mathcal{H} \otimes \mathcal{H} \otimes \cdots \otimes \mathcal{H}$, where

$$p_I = p_{i_1} p_{i_2} \cdots p_{i_n}, \qquad |\psi_I\rangle = |\psi_{i_1}\rangle \otimes |\psi_{i_2}\rangle \otimes \cdots \otimes |\psi_{i_n}\rangle$$

when $I = (i_1, i_2, \dots, i_n) \in \{1, 2, \dots, m\}^n$. The product structure expresses that the generation of the quantum state is a process without memory. The states of different shots are statistically independent.

Now we are ready to define what we mean by a *reliable compression* of a source $(p_i, |\psi_i\rangle)$ on a Hilbert space $\mathcal{H}$. The compression scheme consist of two quantum operations $\mathcal{C}^n : \mathcal{B}(\mathcal{H}^n) \to B(\mathcal{K}_n)$ and $\mathcal{D}^n : \mathcal{B}(\mathcal{K}_n) \to B(\mathcal{H}^n)$. $\mathcal{K}_n$ is a Hilbert space of dimension 2^{nR_n}. We assume that $\mathcal{K}_n \subset \mathcal{H}^n$) and $\mathcal{D}^n(D) = D \oplus 0$. This compression scheme is *reliable* and has *rate* R when

(i) $R_n \to R$

(ii) $\sum_I p_I F\big(|\psi_I\rangle\langle\psi_I|, \mathcal{D}^n \cdot \mathcal{C}^n\big(|\psi_I\rangle\langle\psi_I|\big)\big) \to 1$ as $n \to \infty$.

The first condition tells that asymptotically 2^R dimension is used for the compression of a single emission of the source on the average. (This dimension is equivalent to the use of R qubits) On the other hand, the second condition tells that the emitted state and the compressed one are close in the average, the expectation value of the fidelity is converging to 1. (Note that this definition of the reliable compression scheme is not the most general, since the form of $\mathcal{D}^n$ is rather restrictive.)

The key role of the quantum extension of Shannon's first theorem is played by the *von Neumann entropy.* If D is a density matrix, then its eigenvalues $\lambda_1, \lambda_2, \dots, \lambda_k$ are nonnegative and von Neumann set

$$S(D) := -\sum_i \lambda_i \log \lambda_i. \tag{3}$$

Concerning the von Neumann entropy, see the Appendix.

The positive part of *Schumacher's source coding theorem* is the following.

Theorem 4. *Let $(p_i, |\psi_i\rangle)$ be a source of pure states on a Hilbert space $\mathcal{H}$ and let S be the von Neumann entropy of the density matrix $\sum_i p_i |\psi_i\rangle\langle\psi_i|$. If $R > S$, then there exists a reliable compression scheme of rate R.*

Proof. Let $\xi_1, \xi_2, \dots, \xi_k$ be the eigenvectors of the density matrix $\sum p_i |\psi_i\rangle\langle\psi_i|$ and $\lambda_1, \lambda_2, \dots, \lambda_k$ be the corresponding eigenvalues. $(\lambda_1, \lambda_2, \dots, \lambda_k)$ is a probability distribution on the set $\mathcal{X} := \{1, 2, \dots, k\}$ and $H(\lambda_1, \lambda_2, \dots, \lambda_k) = S$. We shall use the universal coding method of Csiszár and Körner (see the proof of Theorem 3).

Let

$$\mathcal{P}_n^\circ = \{P \in \mathcal{P}_n : H(P) \le R_n\},$$

where $R_n = R - \dfrac{k}{n}\log(n+1)$. We showed above that for

$$A_n := \{I \in \mathcal{X}^n : P_I \in \mathcal{P}_n^\circ\}$$

we have

$$\#(A_n) \le 2^{nR} \qquad \text{and} \qquad \sum_{I \in A_n} \lambda_I \to 1 \quad \text{as } n \to \infty$$

where $\lambda_I = \lambda_{i(1)}\lambda_{i(2)} \dots \lambda_{i(n)}$ for $I = \big(i(1), i(2), \dots, i(n)\big)$. The Hilbert space $\mathcal{K}_n$ for the compressing scheme will be a subspace of $\mathcal{H}^n$,

$$\{\xi_I : I \in A_n\}$$

is a basis for $\mathcal{K}_n$. We have $\dim \mathcal{K}_n \le 2^{nR}$. Next we give the quantum operations $\mathcal{C}^n : \mathcal{B}(\mathcal{H}^n) \to \mathcal{B}(\mathcal{K}_n)$ and $\mathcal{D}^n : \mathcal{B}(\mathcal{K}_n) \to \mathcal{B}(\mathcal{H}^n)$. Set

$$\mathcal{C}^n(\sigma) = P_n \sigma P_n + \sum_{I \notin A_n} A_I \sigma A_I^*$$

where P_n is the orthogonal projection $\mathcal{H}^n \to \mathcal{K}_n$, $A_I = |\xi\rangle\langle\xi_I|$ with a fixed vector $\xi \in \mathcal{K}_n$. For $\rho \in \mathcal{B}(\mathcal{K}_n)$ $\mathcal{D}^n(\rho)$ acts on $\mathcal{K}_n \subset \mathcal{H}^n$ and ρ and 0 on $\mathcal{H}^n \ominus \mathcal{K}_n$. Our task is to show that

$$F_n := \sum_I p_I F\big(|\psi_I\rangle\langle\varphi_I|,\ \mathcal{D}^n \cdot \mathcal{C}^n\big(|\psi_I\rangle\langle\psi_I|\big)$$

converges to 1. We give a lower estimate simply by neglecting the second term in the definition of $\mathcal{C}^n(\sigma)$:

$$\begin{aligned} F_n &\ge \sum_{I \in A_n} p_I \|P_n \psi_I\|^2 = \sum_{I \in A_n} p_I \sum_{J \in A_n} |\langle \psi_I \mid \xi_J\rangle|^2 \\ &= \sum_{J \in A_n} \lambda_J \end{aligned}$$

since

$$\sum_I p_I |\langle \psi_I \mid \xi_J \rangle|^2 = \langle \xi_J | D \otimes \dots \otimes D | \xi_J \rangle = \lambda_J .$$

Above we observed that the lower bound goes to 1, hence $F_n \to 1$, in fact exponentially. ■

Note that the pure states $|\psi_i\rangle$ compressed into mixed state in the scheme we have constructed. It is also remarkable that the statistical operator $\sum p_i |\psi_i\rangle\langle\psi_i|$ of the ensemble played a key role and not the ensemble itself. (Many different ensembles may have the same statistical operator.) To construct the compression we used the eigenbasis of the statistical operator and the value of its entropy. No further data was necessary.

The negative part of Schumacher's theorem depends on the *high probability subspace theorem* obtained by Hiai, Ohya and Petz ([16], [11]).

Theorem 5. *Let D be a density matrix acting on the Hilbert space $\mathcal{H}$. Then the n fold tensor product $D_n := D \otimes D \otimes \cdots \otimes D$ acts on the n-fold product space $\mathcal{H}_n := \mathcal{H} \otimes \mathcal{H} \otimes \cdots \otimes \mathcal{H}$. For any $1 > \varepsilon > 0$ we have*

$$\lim_{n\to\infty} \frac{1}{n} \inf\{\log \operatorname{Tr} Q_n : Q_n \text{ is a projection on } \mathcal{H}_n,\ \operatorname{Tr} D_n Q_n \geq 1-\varepsilon\} = S(D).$$

Roughly speaking the theorem tells that a projection Q_n of large probability has the dimension $\exp\big(nS(D)\big)$.

Proof. First we construct projections of high probability and of small dimension. Fix $\delta > 0$ and let $P(n,\delta)$ be the spectral projection of $-\frac{1}{n}\log D_n$ corresponding to the interval $S(D)-\delta,\ S(D)+\delta$. It follows that

$$\big(S(D)-\delta\big)\,P(n,\delta) \leq \left(-\frac{1}{n}\log D_n\right) P(n,\delta) \leq \big(S(D)+\delta\big)\,P(n,\delta)$$

and hence

$$e^{-n(S(D)+\delta)} P_{(n,\delta)} \leq D_n P(n,\delta) \leq e^{-n(S(D)-\delta)} P_{(n,\delta)}.$$

From this we easily conclude

$$\frac{1}{n}\log \operatorname{Tr} P(n,\delta) \leq S(D)+\delta.$$

and $\limsup_{n\to\infty} \leq S(D)$ follows concerning the limit in the statement.

Let now Q_n be a projection on $\mathcal{H}_n$ such that $\operatorname{Tr} Q_n D_n \geq 1-\varepsilon$. This implies

$$\liminf_{n\to\infty} D_n Q_n P(n,\delta) \geq 1-\varepsilon$$

since

$$\begin{aligned} \operatorname{Tr} D_n Q_n P(n,\delta) &= \operatorname{Tr} D_n Q_n - \operatorname{Tr} D_n Q_n P(n,\delta)^{\perp} \\ &\geq \operatorname{Tr} D_n Q_n - \operatorname{Tr} D_n P(n,\delta)^{\perp}. \end{aligned}$$

Next we estimate as follows:

$$\begin{aligned} \operatorname{Tr} Q_n &\geq \operatorname{Tr} Q_n P(n,\delta) \\ &\geq \operatorname{Tr} D_n Q_n P(n,\delta) e^{n(S(D)-\delta)} \\ &= e^{n(S(D)-\delta)} \cdot \operatorname{Tr} D_n Q_n P(n,\delta) \end{aligned}$$

and

$$\frac{1}{n}\log \operatorname{Tr} Q_n \geq S(D)-\delta+\frac{1}{n}\log \operatorname{Tr} D_n Q_n P(n,\delta).$$

When $n\to\infty$ the last term of the right hand side converges to 0. ■

Now we turn back to Schumacher's theorem and present the negative part.

Theorem 6. *Let $(p_i, |\psi_i\rangle)$ be a source of pure states on a Hilbert space $\mathcal{H}$ and let S be the von Neumann entropy of the density matrix $\sum_i p_i |\psi_i\rangle\langle\psi_i|$. If $R < S$, then reliable compression scheme of rate R does not exists.*

Proof. Assume that a reliable compression scheme of rate $R < S$ exists. Then

$$\begin{aligned}
F_n :&= \sum_I p_I F\big(|\psi_I\rangle\langle\varphi_I|\big),\ \mathcal{D}^n \cdot \mathcal{C}^n\big(|\psi_I\rangle\langle\psi_I|\big) \\
&= \sum_I p_I \sqrt{\langle\psi_I|\mathcal{C}^n\big(|\psi_I\rangle\langle\psi_I|\big)\,|\psi_I\rangle} \\
&\le \sqrt{\sum_I p_I \langle\psi_I|\mathcal{C}^n\big(|\psi_I\rangle\langle\psi_I|\big)\,|\psi_I\rangle}
\end{aligned}$$

by concavity. Moreover,

$$\begin{aligned}
\sum_I p_I \langle\psi_I|\mathcal{C}^n\big(|\psi_I\rangle\langle\psi_I|\big)\,|\psi_I\rangle &= \sum_I p_I \operatorname{Tr} |\psi_I\rangle\langle\psi_I| P_n \mathcal{C}^n\big(|\psi_I\rangle\langle\psi_I|\big) \\
&\le \sum_I p_I \operatorname{Tr} |\psi_I\rangle\langle\psi_I| P_n = \operatorname{Tr} D_n P_n
\end{aligned}$$

for the projection P_n of $\mathcal{H}^n$ onto $\mathcal{K}_n$. Since the compression is of rate R we have

$$\lim_n \frac{1}{n} \log \dim P_n \le R.$$

On the other hand, the high probability subspace theorem tells us that in this case

$$\limsup_n \operatorname{Tr} D_n P_n \ge 1 - \varepsilon$$

is impossible for any $0 < \varepsilon < 1$. We have arrived at a contradiction with the assumption that the average fidelity is converging to 1. In fact, we have shown that for any compression scheme of rate R the average fidelity converges to 0. ■

3. Extension to sources with memory

Extension of Schumacher's source coding theorem is possible in several directions. One way would be to allow a source of mixed states. Not much is known about this direction and we refer to [2], where this problem is discussed. Our attention here will be focused on sources having some memory but producing pure states. In this case the optimal compression rate depends on the density matrix of the source only.

Let $\mathcal{H}$ be a finite-dimensional Hilbert space and $\mathcal{H}_n := \mathcal{H} \otimes \mathcal{H} \otimes \cdots \otimes \mathcal{H}$. Let X^n denote the set of all messages of length n. If $\mathbf{x} \in X^n$ is a message, then a quantum state $\big|\psi(\mathbf{x})\big\rangle$ of the n-fold quantum system is corresponded with it. If $\mathbf{x}$ appears with probability $p(\mathbf{x})$, then

$$D_n := \sum_{\mathbf{x}} p(\mathbf{x}) \big|\psi(\mathbf{x})\big\rangle \big\langle \psi(\mathbf{x})\big|$$

is the density matrix of the n-shot source. This general formulation allows the case

$$\big|\psi(\mathbf{x})\big\rangle = \big|\psi(x_1)\big\rangle \otimes \big|\psi(x_2)\big\rangle \otimes \ldots \big|\psi(x_n)\big\rangle \qquad (\mathbf{x} = (x_1, x_2, \ldots, x_n),$$

but the scheme is more general. It turns out that the optimal compression rate will depend on the density matrices D_n only, hence we do not assume anything about the probability distributions $p(\mathbf{x})$, however we make some assumption on the sequence D_n of density matrices. We always assume that the von Neumann entropy density

$$h := \lim_{n \to \infty} \frac{1}{n} S(D_n) \tag{4}$$

exists. This holds in many examples. For $0 < \varepsilon < 1$, set

$$\beta_n(\varepsilon) := \inf \big\{ \log \mathrm{Tr}(q)) \ : q \text{ is a projection on } \mathcal{H}_n, \mathrm{Tr}\, D_n q \geq 1 - \varepsilon \big\}.$$

We shall say that the high-probability subspace theorem holds if

$$\textbf{(HP)} \qquad \lim_{n \to \infty} \frac{1}{n} \beta_n(\varepsilon) = h.$$

Since we want to let $n \to \infty$, it is reasonable to view all the n-fold systems as subsystems of an infinite one. Let an infinitely extended system be considered over the lattice $\mathbb{Z}$ of integers. The observables confined to a lattice site $k \in \mathbb{Z}$ form the self-adjoint part of a finite-dimensional matrix algebra $\mathcal{A}_k$, that is the set of all operators acting on the finite-dimensional space $\mathcal{H}$. It is assumed that the local observables in any finite subset $\Lambda \subset \mathbb{Z}$ are those of the finite quantum system

$$\mathcal{A}_\Lambda = \bigotimes_{k\in\Lambda} \mathcal{A}_k.$$

The quasilocal algebra $\mathcal{A}$ is the norm completion of the normed algebra $\mathcal{A}_\infty = \cup_\Lambda \mathcal{A}_\Lambda$, the union of all local algebras $\mathcal{A}_\Lambda$ associated with finite intervals $\Lambda \subset \mathbb{Z}$.

A state φ of the infinite system is a positive normalised functional $\mathcal{A} \to \mathbb{C}$. It does not make sense to associate a statistical operator to a state of the infinite system in general. However, φ restricted to a finite-dimensional local algebra $\mathcal{A}_\Lambda$ admits a density matrix D_Λ. We regard the algebra $\mathcal{A}_{[1,n]}$ as the set of all operators acting on the n-fold tensor product space $\mathcal{H}^{\otimes n}$. Moreover, we assume that the density D_n from the first part of this section is identical with $D_{[1,n]}$. Under this assumptions we call the state φ the state of the (infinite) channel. If φ happens to be a product, then we are in the memoryless setting discussed above. Now we want to allow memory effect and pose weaker conditions.

The right shift on the set $\mathbb{Z}$ induces a transformation γ on $\mathcal{A}$. A state φ is called *stationary* if $\varphi \circ \gamma = \varphi$. The state φ is called *ergodic* if it is an extremal point in the set of stationary states. Moreover, φ is *completely ergodic* when it is an extreme point for every $m \in \mathbb{N}$ in the convex set of all states ψ such that $\psi \circ \gamma^m = \psi$.

A weaker form of property **HP** was proven in [11] for a completely ergodic stationary state. The weak high-probability subspace theorem holds if

(w-HP)

$$\limsup_{n\to\infty} \frac{1}{n}\beta_n(\varepsilon) \le h \text{ and } \liminf_{n\to\infty} \frac{1}{n}\beta_n(\varepsilon) \ge \frac{1}{1-\varepsilon}h - \frac{\varepsilon}{1-\varepsilon}\log d,$$

and the McMillan-type convergence holds if

(McM) $$\lim_{n\to\infty} \frac{1}{n}\log D_n = h\cdot I$$

in a certain topology. Loosely speaking **McM** $\Longrightarrow$ **HP** $\Longrightarrow$ **w-HP** and all these properties imply that the extension of Schumacher's theorem holds and the optimal compression rate is the von Neumann entropy density h.

A nice class of stationary states is formed by the *quantum Markov states* (finitely correlated states, algebraic states, or generalised Markov chains are other names for the same thing, see [1, 7, 12]). For those states property **HP** was also proved [12].

Let $H_{[1,n]}$ be the Hamiltonian of *stationary interaction of finite range,* see [3], or Sect. 15 of [16] for the definitions. Assume that D_n is the density of the local Gibbs state, that is

$$D_n := \frac{e^{H_{[1,n]}}}{\operatorname{Tr} e^{H_{[1,n]}}}$$

and let ψ be the equilibrium state of the infinite system. The equilibrium state of the finite system $\mathcal{H}^{\otimes n}$ with some specified interaction could be a model for storing information. If the interaction is stationary and of finite range then the asymptotically optimal compression rate is again the von Neumann entropy density h, since the **McM** property holds in the GNS space for the state ψ [10].

4. Appendix

4.1. Stochastic mappings as state transformations

Assume that $\mathcal{H}$ is the Hilbert space of our quantum system which initially has a statistical operator D (acting on $\mathcal{H}$). When the quantum systems is not closed, it is coupled to another system, called *environment.* The environment has a Hilbert space $\mathcal{H}_e$ and statistical operator D_e. Before interaction the total system has density $D_e \otimes D$. The dynamical change caused by the interaction is implemented by a unitary and $U(D_e \otimes D)U^*$ is the new statistical operator and the reduced density $\tilde{D}$ is the new statistical operator of the quantum system we are interested in. The affine change $D \mapsto \tilde{D}$ is typical for quantum mechanics and called *stochastic mapping.*

The above defined stochastic mapping can be described in several other forms, reference to the environment could be omitted completely. Assume that D is an $n \times n$ matrix and D_e is of the form $(z_k \overline{z_l})_{kl}$ where $(z_1, z_2, \dots, z_m)$ is a unit vector in the m dimensional space $\mathcal{H}_e$. (D_e is pure state.) All operators acting on $\mathcal{H}_e \otimes \mathcal{H}$ are written in a block matrix form, they are

$m \times m$ matrices with $n \times n$ matrix entries. In particular, $U = (U_{pq})_{p,q=1}^{m}$ and $U_{pq} \in M_n$. The definition of the reduced density matrix gives

$$\begin{aligned}\tilde{D} &= \sum_{p,q,r} U_{pq}(D_e \otimes D)_{qr}(U^*)_{rp} = \sum_p \Big(\sum_q z_q U_{pq}\Big) D \Big(\sum_r z_r U_{pr}\Big)^* \\ &= \sum_p A_p D A_p^*\end{aligned}$$

where the operators $A_p := \sum_q z_q U_{pq}$ satisfy

$$\sum_p A_p A_p^* = I. \tag{5}$$

Theorem 7. *Any stochastical mapping $D \mapsto \tilde{D}$ can be written in the form*

$$\tilde{D} = \sum_p A_p D A_p^*,$$

where the operator coefficients satisfy (5). *Conversely, all transformation of this form are stochastic.*

The first part of the theorem was obtained above. To prove the converse part, we need to solve the equations

$$\sum_q z_q U_{pq} = A_p \qquad (p = 1, 2, \ldots, m).$$

Choose simply $z_1 = 1$ and $z_2 = z_3 = \cdots = z_m = 0$ and the equations reduce to $U_{p1} = A_p$. This means that the first column is given from the block matrix U and we need to determine the other columns such a way that U should be a unitary. Thanks to the condition (5) this is possible. Condition (5) tells us that the first column of our block matrix determines an isometry which extends to a unitary. ■

The coefficients A_p in the *operator-sum representation* are called the *operation elements* of the stochastic map. The term quantum (state) operation is also often used instead of stochastic map.

The stochastic maps form a convex subset of the set of all positive trace preserving linear transformations.

4.2. Von Neumann entropy

The above formula for the von Neumann entropy is equivalently written as

$$S(D) = \operatorname{Tr} \eta(D), \quad \text{where} \quad \eta(t) = -t \log t.$$

Since η is a concave function on $\mathbb{R}^+$, the von Neumann entropy is a *concave functional* on the state space. The maximum is reached at the density whose all the eigenvalues are the same and the minimum is at pure states.

A density matrix D admits generally many convex decomposition into pure states: $D = \sum_j \mu_j |\psi_j\rangle\langle\psi_j|$. The von Neumann entropy is the infimum of all Shannon entropies corresponding to those decompositions

$$S(D) = \inf \Big\{ H(\mu_j) \,:\, D = \sum_j \mu_j |\psi_j\rangle\langle\psi_j|, \mu_j \geq 0, \sum_j \mu_j = 1 \Big\}.$$

When D_{12} is a density matrix of a composite system $\mathcal{H}_1 \otimes \mathcal{H}_2$ with reduced density matrices D_1 and D_2, respectively, then the *subadditivity* of the von Neumann entropy holds:

$$S(D_{12}) \leq S(D_1) + S(D_2)$$

This property is responsible for the fact that for shift invariant states of an infinite tensor product the von Neumann entropy density (4) exists.

Several chapters of the book [16] are devoted to properties and extensions of the von Neumann entropy. For a historical approach to the von Neumann entropy, see [17].

4.3. Fidelity

The general formula for the fidelity of the density matrices D_1 and D_2 is

$$F(D_1, D_2) = \operatorname{Tr} \sqrt{D_1^{1/2} D_2 D_1^{1/2}}. \tag{6}$$

This quantity was studied by Uhlmann in a different context and he proved that

$$F(D_1, D_2) = \min \big\{ \sqrt{\operatorname{Tr}(D_1 G)\, \operatorname{Tr}(D_2 G^{-1})} \,:\, G \text{ is positive and invertible} \big\} \tag{7}$$

([20] and see [8] for a rather detailed discussion). From this the symmetry of $F(D_1, D_2)$ is obvious and we can easily deduce the *monotonicity of the fidelity* under stochastic state transformation:

$$\begin{aligned}F\big(\mathcal{C}(D_1),\mathcal{C}(D_2)\big)^2 &\ge \operatorname{Tr}\mathcal{C}(D_1)G \operatorname{Tr}\mathcal{C}(D_2)G^{-1} - \varepsilon \\ &\ge \operatorname{Tr} D_1\mathcal{C}^*(G) \operatorname{Tr} D_2\mathcal{C}^*(G^{-1}) - \varepsilon \\ &\ge \operatorname{Tr} D_1\mathcal{C}^*(G) \operatorname{Tr} D_2\mathcal{C}^*(G)^{-1} - \varepsilon \\ &\ge F(D_1, D_2)^2 - \varepsilon,\end{aligned}$$

where $\mathcal{C}^*$ is the adjoint of $\mathcal{C}$ with respect to the Hilbert-Schmidt inner product, $\varepsilon > 0$ is arbitrary and G is chosen to be appropriate. It is well-known that $\mathcal{C}^*$ is unital and positive, hence $\mathcal{C}^*(G)^{-1} \ge \mathcal{C}^*(G^{-1})$. In this way the monotonicity

$$F\big(\mathcal{C}(D_1),\mathcal{C}(D_2)\big) \ge F(D_1, D_2) \tag{8}$$

is concluded.

From the definition (6) one observes that $F(D_1, D_2)$ is concave in D_2. (Remember that $\sqrt{t}$ is operator concave.) However, the monotonicity gives that $F(D_1, D_2)$ is *jointly concave* as well. Consider the stochastic mapping

$$\mathcal{C} : \begin{bmatrix} A & B \\ C & D \end{bmatrix} \mapsto A + D$$

Then

$$\begin{aligned}&\lambda F(D_1, D_2) + (1-\lambda)F(D_1', D_2') \\ &= F\left(\begin{bmatrix}\lambda D_1 & 0 \\ 0 & (1-\lambda)D_1'\end{bmatrix}, \begin{bmatrix}\lambda D_2 & 0 \\ 0 & (1-\lambda)D_2'\end{bmatrix}\right) \\ &\le F\big(\lambda D_1 + (1-\lambda)D_1', \lambda D_2 + (1-\lambda)D_2'\big)\end{aligned}$$

as an application of monotonicity and the concavity is obtained.

Another remarkable formula is

$$\begin{aligned}F(D_1, D_2) = \max\Big\{\big|\langle\psi_1|\psi_2\rangle\big| \ :\ &\mathcal{C}\big(|\psi_1\rangle\langle\psi_1|\big) = D_1, \\ &\mathcal{C}\big(|\psi_2\rangle\langle\psi_2|\big) = D_2 \text{ for some stochastic mapping } \mathcal{C}\Big\}\end{aligned}$$

which has a certain *operational meaning* [6].

REFERENCES

[1] L. Accardi and A. Frigerio, Markov cocycles, *Proc. R. Ir. Acad.,* **83A** (1983), 251–269.

[2] H. Barnum, C. M. Caves, C. A. Fuchs, R. Jozsa and B. W. Schumacher, On quantum coding for ensembles of mixed states, *J. Phys. A,* **34** (2001), 6767–6785.

[3] O. Bratteli and D. W. Robinson, *Operator Algebras and Quantum Statistical Mechanics II,* Springer-Verlag, New York–Heidelberg–Berlin (1981).

[4] T. M. Cover and J. A. Thomas, *Elements of information theory,* Wiley (1991).

[5] I. Csiszár and J. Körner, *Information theory. Coding theorems for discrete memoryless systems,* Akadémiai Kiadó, Budapest (1981).

[6] J. L. Dold and M. A. Nielsen, A simple operational interpretation of fidelity, arXiv e-print quant-ph/0111053.

[7] M. Fannes, B. Nachtergaele and R. F. Werner, Finitely correlated states on quantum spin chains, *Comm. Math. Phys.,* **144** (1992), 443–490.

[8] C. A. Fuchs, *Distinguishability and Accessible Information in Quantum Theory,* The university of New Mexico, Ph.D. thesis, Albuquerque (1996), arXiv e-print quant-ph/9601020.

[9] R. M. Gray, *Entropy and Information Theory,* Springer (1990).

[10] F. Hiai, M. Ohya, D. Petz, McMillan type convergence for quantum Gibbs states, *Arch. der Math.,* **65** (1995), 154–158.

[11] F. Hiai and D. Petz, The proper formula for relative entropy and its asymptotics in quantum probability, *Commun. Math. Phys.,* **143** (1991), 99–114.

[12] F. Hiai and D. Petz, Entropy density for algebraic states, *J. Functional Anal.,* **125** (1994), 287–308.

[13] A. S. Holevo, Quantum coding theorems, *Russian Math. Surveys,* **53** (1998), 1295–1331.

[14] R. Jozsa and B. Schumacher, A new proof of the quantum noiseless coding theorem, *J. Modern Optics,* **41** (1994), 2343–2349.

[15] M. A. Nielsen and I. L. Chuang, *Quantum Computation and Quantum Information,* Cambridge University Press (2000).

[16] M. Ohya and D. Petz, *Quantum Entropy and Its Use,* Springer-Verlag, Heidelberg (1993).

[17] D. Petz, Entropy, von Neumann and the von Neumann entropy, in: *John von Neumann and the Foundations of Quantum Physics,* eds. M. Rédei and M. Stöltzner, Kluwer (2001).

[18] D. Petz and M. Mosonyi, Stationary quantum source coding, *J. Math. Phys.,* **42** (2001), 4857–4864.

[19] B. Schumacher, Quantum coding, *Phys. Rev. A,* **51** (1995), 2738–2747.

[20] A. Uhlmann, *Rep. Math. Phys.,* **9** (1976), 273.

Dénes Petz
Department for Mathematical Analysis
Budapest University of Technology and Economics
H-1521 Budapest XI.
Hungary

petz@renyi.hu

BOLYAI SOCIETY
MATHEMATICAL STUDIES, 16

Entropy, Search, Complexity, pp. 179–207.

Information Theory at the Service of Science

FLEMMING TOPSØE*

Information theory is becoming more and more important for many fields. This is true for engineering- and technology-based areas but also for more theoretically oriented sciences such as probability and statistics.

Aspects of this development is first discussed at the non-technical level with emphasis on the role of information theoretical games. The overall rationale is explained and central types of examples presented where the game theoretical approach is useful.

The final section contains full proofs related to a subject of central importance for statistics, the estimation or updating by a posterior distribution which aims at minimizing divergence measured relative to a given prior.

1. Introduction, background

Information Theory is of importance for a number of disciplines from the very applied engineering oriented ones to more theoretical fields. One of the strongest interfaces is to probability and statistics. One can see a line in monographs such as Čencov [10], Kullback [43], Dembo and Zeituni [23] and Cover and Thomas [11], and in recent and ongoing research of authors including Amari [1], Johnson and Barron [5], Csiszár and Matúš [19], Harremoës [29] and others, much of it in the publishing process. General purpose textbooks which aim at making the new tools available are starting to appear, though still not taking full advantage of recent improvements and extensions of the techniques (cf. Jessop [38] and Applebaum [2]).

*Research supported by INTAS, project 00-738, and by the Danish Natural Science Research Council.

Information theory offers technical tools and a conceptual base which contribute to an understanding of many of the fundamental concepts and methods of probability and statistics. This also leads to a better understanding of the basic notion of probability. In this connection recall what de Finetti wrote in the foreword to his monograph [22] after a life long study of these fields: "Probabilities do not exist!" That there is something strange about the concept of probability is nowadays realized by all workers in theoretical probability and statistics and the last word on the concept of probability, on randomness, has certainly not been said. Is it something inherent in the real world or only something going on in our head, in our *perception* of the real world?

The change of paradigm which appears to be underway may be based on classical probability theory "à la Kolmogorov". However, more drastic changes may be underway, either based on information theory proper or on the neighbouring and inter-related discipline of complexity theory. In this connection we refer to Shafer and Vovk [50] which, at a broader level, is indicative of what may be coming. For important contributions aiming more directly at information theory, we point to a series of papers by Rissanen which can be traced from the survey article by Barron, Rissanen and Yu [48].

It has to be recognized that statistical thinking – also dating back to "pre-information theory times" – contains ideas and concepts which are in line with information theoretical thinking. Thus, it is perhaps more appropriate to say that one can expect a much stronger cross-fertilization between the two fields of information theory and statistics than seen before.

In order to understand what information theory has to offer which could influence basic probabilistic and statistical thinking, we point to three factors: The basic concepts have natural interpretations and as such are "just the right ones" and, secondly, you can often use these concepts to model *conflict situations* which are common in many areas, such as biology, economics, physics and then also in probability and, more pronouncedly so, statistics. As a last reason we point to the technical tools of information theory, especially to the powerful inequalities.

As to the first feature pointed to, we owe to Shannon, cf. his pioneering article [51] from 1948, the definition of "the right concepts" such as entropy, conditional entropy, mutual information and so on. Especially important for the interface to statistics is the introduction in 1951 by Kullback and Leibler [44], of a further quantity, now mainly called (*information*) *divergence*, cf. also Kullback [43].

The ability of information theory to model conflict situations can be said to go back to Shannon himself. As an early source we also mention Kelly [41]. However, we want to emphasize the comprehensive study of the Maximum Entropy Principle (MaxEnt) by Jaynes who from 1957 onwards has worked to put this principle on a firm information theoretical basis and also discussed the qualitative and philosophical aspects at great length, cf. [35], [36], [34] and [37]. MaxEnt dictates that if $\mathcal{P}$ is a set consisting of probability distributions, then the least biased one, and hence the one best suited for predictions, is that distribution in the set – if such a unique distribution exists – which has maximal entropy. Jaynes further stressed the view – which is at the same time a guiding principle – that one should not think of the set as a set of distributions, one of which is the "true" distribution. Rather, Jaynes maintains that $\mathcal{P}$ models our *knowledge* about the system studied.

Some years ago the author pointed out that a principle of *Game Theoretical Equilibrium* (GTE) can be taken to be basic, cf. [56] and [57]. Thus, from GTE, you are led to MaxEnt as well as to a principle going back to Kullback, the *Minimum Discrimination Information Principle,* which dictates that you select that distribution in the model which has the smallest divergence to a given or suitably chosen *prior* distribution. As the term "divergence" now appears to be more common than "discrimination" (or "information distance" or "relative entropy" which are also terms in use), we refer to Kullback's principle as MinDiv, the *minimum divergence principle.* The geometrically-flavoured structure involved in MinDiv was studied by Čencov [9] and by Csiszár [14], who introduced the concept of *I-projection* for the optimal distribution of MinDiv. Further studies of Csiszár demonstrated the significance of information theory for statistics, cf. [15].

The success of information theory at the service of other sciences depends on the ability to derive key results based on information theoretical principles in a way which is felt natural and technically convenient within the sciences in question. The sources mentioned contain results in this direction for the fields of probability and statistics. Some additional references include the difficult paper by Linnik, [45], regarding the Central Limit Theorem, Barron's follow-up paper, [4], and results to come by Barron and Johnson, [5], and we may also mention Harremoës [27] and Topsøe [59]. The paper [26] by Grünwald and Dawid is very much in line with the approach adopted here. The reader should also watch out for forthcoming work by Boyd and Chiang.

2. Game Theoretical Equilibrium, the idea

In this section we shall reveal the over all idea of GTE. In principle, this is simple: GTE dictates that you use information theoretical concepts to view a problem as a game, typically as a two-person zero-sum game. The search for optimal strategies, i.e. the attempt to approach equilibrium in the game theoretical sense, will then lead to useful, even completely satisfactory insight into the original problem. The power of GTE then depends on the ability to make meaningful transformations of interesting problems to the game theoretical setting. Below follow some qualitative, partly philosophical remarks on this issue.

Information theory provides concepts and tools for expressing the role of an "observer" or "decision maker" (the statistician, the physicist, the investment planner or what the case may be). As a brief indication of this aspect of information theory, we need only point to coding as a way to *represent* or to *describe* data and, at the same time, as a means to *identify* outcomes, hence to express strategies for making *observations.*

When we use information theory to model a part of *"reality"*, we should acknowledge – in consistency with a Platonic or neo-Platonic view – that all we can do is to model our *ignorance* about reality or, put positively, to model our *knowledge* about reality. In our modelling we should be open to any eventuality. Possibly motivated by the experience that "what can go wrong *does* go wrong", this points to applying a safe (cowardice!) strategy of minimizing maximal risk, hence this leads to by now classical "mini-max thinking" of game theory. In its simplest form, we are thus led to consider two-person zero-sum games. This modelling is possible in many specific situations via the identification of an *objective function* (a *cost* or a *pay-off,* depending on the sign). For a discussion of concepts of game theory, in particular the various notions of strategy, see Straffin [52].

Qualitatively, the above considerations are in consistency with basic Bayesian thinking in statistics. And other sciences acknowledge similar ideas. In physics we point to the Copenhagen interpretation of quantum physics and the emphasis of Niels Bohr on the interplay, *complementarity,* between the system under study and the observer looking at it through whatever glasses are available. A system behaves not only in accordance with what is known because of insight gained by previous observations. The behaviour depends on *all* factors – including those we *could* have taken into account. Regarding quantum physics it may well be that the Copenhagen interpretation is ripe for replacement or thorough revision, largely because the notion of probability is not well explained and it is desirable with a change which would involve, among other elements, Jaynes' ideas which combine information theory and the interpretation of probability concepts. For the field of biology, we point to Maynard Smith's principle of *evolutionary stability.* For a recent application of this principle, see Broom [8]. We may also point to the social sciences, to economy, cf. von Neumann's pioneering work in [60], or more modern texts such as Aubin [3], and also, we can point to recent applications to finance where Delbaen et al. [21] and Bellini and Frittelli [6] serve as entrance points to the relevant literature where information theoretical considerations come into play.

3. Information theoretical concepts

3.1. The general idea

In order to stress the underlying rationale, thereby pointing to the wider applicability of basic concepts from information theory, we introduce these concepts in a *context* which is quite abstract and freed from any reference to standard information theory. However, we do use terminology borrowed from that theory.

The nature of the context does not really concern us here. Anyhow, it may be helpful to think of it as some well defined part of "reality" which may have been isolated by some process of *preparation.* As an indication of what we have in mind, consider the case of an ideal gas submerged in a heat bath. Through the preparation, a "system" or "context" suitable for study emerges.

We shall characterize any specific context by three entities. Firstly a set $\mathcal{P}$, called the *knowledge base.* This set reflects the available knowledge. Elements of $\mathcal{P}$ are referred to as *consistent instances,* i.e. as instances which are consistent with the preparation of the system. In specific situations when we know the nature of the elements of $\mathcal{P}$, we may reflect that in our terminology. For instance, in all of our examples, $\mathcal{P}$ will be a set of probability distributions and we may then talk about *consistent distributions.*

Further, we assume that we have access to a set K of *description strategies.* Various interpretations are possible. Either the strategies can be used for making observations from the system or for *representing* consistent instances or, in the terminology we shall prefer below, for *describing* such instances.

The third entity we have in mind is a function, the *objective function,* which to $\kappa \in K$ and $P \in \mathcal{P}$ associates a non-negative (possibly infinite) number, denoted $\langle \kappa, P \rangle$. This is the *description length* (with κ as description strategy and P as the consistent instance).

In the applications we have in mind (see below) the "instances" do not represent final results after observation but rather individual probability distributions. Therefore, to avoid confusion, the reader should note that for such applications, what we refer to as "description length" really represents *mean description length.*

In our set-up we have focused on three elements: Our *knowledge,* the available *tools for description* and our *objective.* Further elements – such as side information, conditioning, symmetry etc. – may be brought into the picture but the chosen framework appears to constitute an adequate playground for basic information theoretical considerations. Instead of the neutral reference to a context, one could refer to the triple $\big(\mathcal{P}, K, \langle \cdot, \cdot \rangle\big)$ as an *information space.*

Sometimes (see Section 5), it is not possible to suggest in a meaningful way an objective function which is non-negative. Often, one can then instead work with an objective function which can also assume negative values. This will, typically, correspond to situations where you find it natural to measure performance relative to some chosen reference. One may, therefore, work with two notions, one of *absolute information spaces* as considered up to now and one of *relative information spaces.* It is only the former which we have in mind in this section.

Now then, let us demonstrate that the set-up as introduced allows the introduction of key quantities for further study.

First define the *entropy* of a consistent instance P as the minimum description length:

$$H(P) = \inf_{\kappa \in K} \langle \kappa, P \rangle. \tag{1}$$

If there exists $\kappa \in K$ with $\langle \kappa, P \rangle = H(P)$ we say that κ is *adapted* to P. Often, such strategies are uniquely determined from P.

Then we can define *redundancy* $D(P \parallel \kappa)$ as the unnecessary part of the description length when using $\kappa \in K$ as strategy instead of an optimal strategy adapted to $P \in \mathcal{P}$, i.e.

$$D(P \parallel \kappa) = \langle \kappa, P \rangle - H(P). \tag{2}$$

This quantity may also be conceived as the saving in description length inherent in the information obtained if we are told what P is. For this interpretation it is understood that the strategy κ is used for description of P before we are told what P is and that we switch to a strategy best adapted to P as soon as P is revealed to us.

Note that (2) may not be adequate as a definition in all cases. Sometimes, as in the case of information theory, it is possible to suggest a meaningful definition which also covers the indeterminate case when (2) leads to the form $D(P \parallel \kappa) = \infty - \infty$. Anyhow, the *linking identity*

$$\langle \kappa, P \rangle = H(P) + D(P \parallel \kappa) \tag{3}$$

always holds (as we agree that the right hand side is $+\infty$ when $H(P) = \infty$).

A further quantity of interest is the *description gain* inherent in a change of strategy from κ to ρ in the description of P. This quantity can even be taken to be more fundamental than entropy as well as redundancy and it is to be expected that it will play a significant role in future research (it already does so but in a less dominant and more implicit way). A natural defining relation would be

$$D(P \parallel \kappa \rightsquigarrow \rho) = \langle \kappa, P \rangle - \langle \rho, P \rangle.$$

However, this is very likely to lead to the indeterminate quantity $\infty - \infty$ and it is better to exploit the linking identity (3) for $\langle \kappa, P \rangle$ as well as for $\langle \rho, P \rangle$ and use the relation

$$D(P \parallel \kappa \rightsquigarrow \rho) = D(P \parallel \kappa) - D(P \parallel \rho) \tag{4}$$

as definition. Of course, this quantity may be negative or even $-\infty$. Normally, situations with $D(P \parallel \kappa \rightsquigarrow \rho) \geq 0$, i.e. with a genuine gain, will have our main interest and other cases can be ignored. From (4) we obtain the relation

$$D(P \parallel \kappa) = D(P \parallel \kappa \rightsquigarrow \rho) + D(P \parallel \rho), \tag{5}$$

which could be called the *linking identity of the second kind* (then (3) is of the first kind). However – unlike (3) – (5) is not necessarily valid in all cases but does hold, e.g. whenever $D(P \parallel \rho) < \infty$.

Further concepts may be introduced but this will, typically, require extra structure, and we shall only look into that in the context of information theory proper.

3.2. Proper information theoretical concepts

Information theory provides the key example where the above definitions apply and have natural interpretations as suggested by the terminology used. Let us briefly go through this. In so doing we shall here restrict the scope and only have the discrete case in mind. In the more technical final section we extend the scope to encompass also the continuous case.

As starting point we shall then take a discrete set $\mathbb{A}$, i.e. a finite or countably infinite set. This is the *alphabet.* As $\mathcal{P}$ we take a set of probability distributions on $\mathbb{A}$ and as K the set of *codes* or, more precisely, *idealized code length functions.* The way we think about $\kappa \in K$ is as an allocation of code words to elements in $\mathbb{A}$, however only paying attention to the lengths of the code words. More precisely, $\kappa \in K$ is a map which to each $i \in \mathbb{A}$ associates a number $\kappa(i) \in [0, \infty]$ in such a way that Krafts equality in idealized form, i.e. $\sum \exp\big(-\kappa(i)\big) = 1$, holds (the idealization refers to the acceptance of arbitrary real numbers as values of κ, and to the choice of the natural base rather than the base 2 for the exponentiation). The interpretations related to this definition are well known, see e.g. [59].

There is an important natural $1-1$ correspondance, written $\kappa \leftrightarrow P$, between codes and distributions for which $P(i) = \exp\big(-\kappa(i)\big)$, $\kappa(i) = -\log P(i)$ (we use "log" to denote natural logarithms).

In order for the definitions of Section 3.1 to apply we also have to specify the objective function $(\kappa, P) \curvearrowright \langle \kappa, P \rangle$. For this we take expected code length, i.e. we put

$$\langle \kappa, P \rangle = \sum \kappa(i) P(i).$$

The definitions of entropy, redundancy and description gain then make good sense. As is classical, $H(P) = \langle \kappa, P \rangle$ with $\kappa \leftrightarrow P$.

We use the correspondance between codes and distributions to define *divergence* (from P to Q) by

$$D(P \parallel Q) = D(P \parallel \kappa) \tag{6}$$

with $\kappa \leftrightarrow Q$ and also to define the following quantity:

$$D(P \parallel Q \rightsquigarrow R) = D(P \parallel \kappa \rightsquigarrow \rho) \tag{7}$$

with $\kappa \leftrightarrow Q$ and $\rho \leftrightarrow R$. Note that $D(P \parallel Q) = D(P \parallel Q \rightsquigarrow P)$. We can think of $D(P \parallel Q \rightsquigarrow R)$ as reflecting an estimation- or prediction- or updating situation with Q as *prior* and R as *posterior distribution.* We may, therefore, call $D(P \parallel Q \rightsquigarrow R)$ the *estimation gain* or the *updating gain* associated with this situation.

Much clarity results by realizing that redundancy (related to description via coding) conceptually preceedes divergence (related to distributions).

It is well known that whereas entropy can only be finite for essentially discrete distributions, redundancy and divergence has a much wider scope of applicability which matches the requirement in probability and statistics to encompass basic continuous distributions. Indeed, in the case of divergence, one is led to the well known formula

$$D(P \parallel Q) = \int \log \frac{dP}{dQ}\, dP \tag{8}$$

(when $P \ll Q$), which may be used as a definition which then covers all eventualities (with $D(P \parallel Q) = \infty$ in case $P \not\ll Q$). Technical details can be found, e.g. in [58] where you will also find important extensions of concepts introduced, especially involving conditioning and datareduction.

4. Instances of GTE

The conceptual base developed in the previous section gives rise to a number of specific two-person zero-sum games of which we mention three. The results of the first two subsections are of particular relevance for applications to probability and statistics. Possible applications lead to interesting characterizations of key distributions, often accompanied by appropriate limit theorems. Many such results can be conceived as instances of the GTE principle as discussed here. Kapur [39] contains a great number of such results. We may also mention Haussler [32], Kazakos [40], Harremoës [27] and Topsøe [59] which are based on the game theoretical approach. The first papers using this approach are Pfaffelhuber [47] and Topsøe [56][1]. For fur-

[1] The paper [47] was only discovered by the author shortly before submission of the final manuscript. Apparently the research of Pfaffelhuber was carried out in 1975 whereas

ther work in this direction see Harremoës and Topsøe [30] as well as work by several authors, including Grünwald and Dawid [26].

4.1. The $H_{\max}$-game

The first game is the *maximum entropy game,* the $H_{\max}$-*game,* which is related to entropy and code length in the discrete case and refers to a given set $\mathcal{P}$ of distributions over the alphabet $\mathbb{A}$. One may also refer to this game as the *code length game.* For the $H_{\max}$-game, Player I ("the system") chooses a consistent distribution $P \in \mathcal{P}$ and Player II (the statistician, say) chooses a code $\kappa \in K$. Description length $\langle \kappa, P \rangle$ is taken as the cost, seen from the point of view of Player II.

We define $H_{\max} = H_{\max}(\mathcal{P})$, the *maximum entropy value* as

$$H_{\max} = \sup_{P \in \mathcal{P}} H(P).$$

For $\kappa \in K$, we define $R(\kappa) = R(\kappa \mid \mathcal{P})$, the *risk* associated with κ, to be

$$R(\kappa) = \sup_{P \in \mathcal{P}} \langle \kappa, P \rangle$$

and we put

$$R_{\min} = \inf_{\kappa \in K} R(\kappa).$$

Then $H_{\max} \leq R_{\min}$. If equality holds with a finite value: $H_{\max} = R_{\min} < \infty$, we say that the system is in *equilibrium.* If $\mathcal{P}$ is convex and $H_{\max} < \infty$, it can be shown that the game is, indeed, in *equilibrium.* Therefore, under quite general conditions, *maximum entropy equals minimum risk.* As the risk was defined via description length, we may also say that *maximum entropy equals minimum description length.*

A technically simple result which is very powerful for the applications is that if there exists a *Nash equilibrium code* adapted to $\mathcal{P}$, i.e. – copying concepts from game theory – a code λ such that $R(\lambda) \leq H(Q) < \infty$ with $\lambda \leftrightarrow Q$ and $Q \in \mathcal{P}$, then λ and Q are unique optimal strategies for the players. In particular, Q is the unique maximum entropy distribution. For a quick introduction to these results, see Topsøe [59]. More details are in Harremoës and Topsøe [31].

---

the main work of the author stems from 1976 with the first formal publication in Danish in 1978. In any case, the author shares the introduction of the game theoretical approach with Pfaffelhuber.

4.2. The $D_{\min}$-game

The second game we shall study is the $D_{\min}$-*game* which we here describe in rather qualitative and motivating terms and again only having the discrete case in mind. We do, however, follow-up in Section 5 with concrete technical details which apply to a more general situation, covering also the continuous case.

The setting is as in Section 4.1, except that now we have also given a *reference code* κ_0 or, equivalently, a *prior distribution* P_0 ($\kappa_0 \leftrightarrow P_0$). Again, Player I chooses a $P \in \mathcal{P}$ and Player II chooses a code $\kappa \in K$ or, equivalently, a *posterior distribution* R (with $\kappa \leftrightarrow R$). But now, description gain, $D(P \parallel \kappa_0 \rightsquigarrow \kappa)$ or, equivalently, *estimation* or *updating gain*, $D(P \parallel P_0 \rightsquigarrow R)$, is taken as pay-off seen from the point of view of Player II. According to the zero-sum character which we insist on, $D(P \parallel \kappa_0 \rightsquigarrow \kappa)$ is considered as a cost to Player I.

If Player I chooses $P \in \mathcal{P}$ one readily sees from the linking identity (5) that fixing this distribution, the largest cost which Player I risks is $D(P \parallel \kappa_0)$. Therefore, an optimal strategy for Player I is a $Q \in \mathcal{P}$ for which $D(Q \parallel \kappa_0) = D_{\min}$ with the latter quantity defined by

$$D_{\min} = \inf_{P \in \mathcal{P}} D(P \parallel P_0).$$

The game thus leads to the MinDiv-principle.

In the theoretical discussion of this game, it is advantegeous to consider also the other side of the game. In normal practice there exists λ such that

$$D(P \parallel \kappa_0 \rightsquigarrow \lambda) \geq D_{\min} \tag{9}$$

for every $P \in \mathcal{P}$. With $\kappa_0 \leftrightarrow P_0$ and $\lambda \leftrightarrow Q$, the inequality (9) is equivalent to the celebrated *Pythagorean inequality:*

$$D(P \parallel P_0) \geq D(P \parallel Q) + D_{\min} \tag{10}$$

for all $P \in \mathcal{P}$. For precise statements and technical detail we refer to Section 5. Here, we point out that Q need not be consistent. The inequality was first proved for cases with $\mathcal{P}$ closed in total variation (hence Q must be consistent) by Csiszár [14]. The general result was established in Topsøe, [56]. In Csiszár's terminology, Q is the (generalized) *I-projection of* P_0 *on* $\mathcal{P}$. With an interpretation as indicated above, Q is the *optimal estimator* or the *optimal updating strategy* based on the prior P_0. In the coding terminology, λ is the *optimal adjustment* of κ_0. Note that if $(P_n)_{n\geq 1} \subseteq \mathcal{P}$ satisfies $D(P_n \| P_0) \to D_{\min}$, then $D(P_n \| Q) \to 0$. This is a rather strong type of limit theorem, in particular stronger than setwise convergence of P_n to Q.

4.3. The $I_{\max}$-game

The third instance which we shall mention and where GTE applies is of main interest for information theory and leads to the *capacity–redundancy game* or, for reasons explained below, the *maximum information transmission game,* the *$I_{\max}$-game.* Again, it involves a knowledge base $\mathcal{P}$ of probability distributions but now no prior code or distribution is given. If Player II ("the receiver") can choose a code and redundancy $D(P \| \kappa)$ is taken as cost, we are led to consider *minimal redundancy* defined as

$$R_{\min} = \inf_{\kappa} \sup_{P\in\mathcal{P}} D(P \| \kappa). \tag{11}$$

Clearly, if the other side of the game is taken to involve only $P \in \mathcal{P}$ as permissable strategies for Player I, the game will normally not be interesting as

$$\sup_{P\in\mathcal{P}} \inf_{\kappa} D(P \| \kappa) = 0. \tag{12}$$

However, if we – "à la von Neumann" – consider mixed strategies (i.e. randomization corresponding to convex combinations $\sum \alpha_\nu P_\nu$ of members of $\mathcal{P}$) and choose the associated *average redundancy,* $\sum \alpha_\nu D(P_\nu \| \kappa)$, as the quantity the players should worry about (the objective function), an interesting game is obtained. Note that we choose as objective function a function which does not only depend on the mixed strategy but also on the actual weights (the α_ν's above). It is the simplest to consider the case when $\mathcal{P}$ consists of a finite or countably infinite set of distributions P_ν $(1 \leq \nu \leq n$ or $1 \leq \nu < \infty)$. This game may be interpreted as one involving a *channel* with $\alpha = (\alpha_\nu)$ as input distribution and then the quantity replacing (12) is

$$\sup_{(\alpha_\nu)} \inf_{\kappa} \sum \alpha_\nu D(P_\nu \| \kappa) \tag{13}$$

which is the *capacity* of the channel (see below). The quantity (11) is not affected as divergence has strong convexity properties, in particular, divergence is convex in the first argument (indeed, if the supremum in

(11) is changed to a supremum over (α_ν)'s and $D(P \parallel \kappa)$ is changed to $\sum \alpha_\nu D(P_\nu \parallel \kappa)$, the same quantity is obtained). By applying the principle of GTE one then proves an important result of information theory, the *capacity-redundancy theorem,* which equates capacity with minimum redundancy. This result is closely related to the well known *Kuhn-Tucker conditions* (for the situation considered, cf. (17) and (18) below). Relevant references include Csiszár and Körner, [17, Problem II.3.1], Davisson and Garcia, [20] and Ryabko [49].

Let us go a little more into the details. First we point out the following identity, the *compensation identity,* cf. Theorems 6.1 and 9.1 of [58]. With notation as above, it states that

$$\sum \alpha_\nu D(P_\nu \parallel \kappa) = \sum \alpha_\nu D(P_\nu \parallel \kappa_0) + D(P_0 \parallel \kappa) \tag{14}$$

where $P_0 = \sum \alpha_\nu P_\nu$ and $\kappa_0 \leftrightarrow P_0$. This shows that the infimum in (13) can be identified as

$$I(\alpha) = \sum \alpha_\nu D(P_\nu \parallel \kappa_0). \tag{15}$$

This quantity is the *information transmission rate* associated with the input distribution α. The maximum information transmission rate, the $I_{\max}$-value, is the *capacity* of the channel and this then is the quantity appearing in (13). The intuitive content of (15) is perhaps best understood if the code κ_0 is replaced by the matching distribution P_0: With probability α_ν a "letter" is sent through the channel and this changes the distribution at the output side from the á priori distribution P_0 to P_ν. The *redundancy removed,* hence the *information received* by this change is $D(P_\nu \parallel P_0)$. Thus, the information transmission rate is defined as average information obtained at the output side. By (15) we may write (14) in the form

$$\sum \alpha_\nu D(P_\nu \parallel \kappa) = I(\alpha) + D(P_0 \parallel \kappa) \tag{16}$$

which is a *linking identity,* now of the third kind.[2]

Intuitively it is to be expected that optimal usage of the channel must be achieved for an input distribution α for which the *Kuhn–Tucker conditions*

$$D(P_\nu \parallel P_0) = C \quad \text{for all } \nu \text{ with } \alpha_\nu > 0 \tag{17}$$

$$D(P \parallel P_0) \le C \quad \text{for all } P \in \mathcal{P} \tag{18}$$

[2] Note that in all cases, a *linking identity* is an identity revealing a basic structural relation which involves an objective function for one of the games considered.

are fulfilled with C some constant. The sufficiency of the conditions follows from the compensation identity. As to the necessity (for $\mathcal{P}$ finite), see Topsøe [55].

When (17) and (18) hold, it follows directly from the compensation identity (14) and from the definitions involved that $I(\alpha) = R_{\min} = C$, hence proving the capacity-redundancy theorem.

The analysis of the $I_{\max}$-game applies in many settings with $D(P \parallel Q)$ replaced by other measures of "divergence". A more subtle analysis turns out to be possible for general *$f-$divergences,* cf. Csiszár [13, Theorem 3.2]. But the simple analysis based on the compensation identity (14) is also possible outside the framework discussed above as this identity holds in a variety of cases. To comment on this, first note that (14) holds for squared Euclidean distance. In that case, the problem suggested by (11) belongs to *location theory,* cf. [25]. The specific problem goes back to Sylvester [53] who wrote "It is required to find the least circle which shall contain a given system of points in a plane" – in fact, this is the full text of [53]!

The compensation identity also holds for *Bregman divergences,* (regarding these, see Bregman [7] and Csiszár [16]).[3]

As a final example of the wide applicability of the Kuhn-Tucker criterion for variants of the $I_{\max}$-game, we mention the analogue of this game in the setting of quantum information theory with $\mathrm{Trace}\, P(\log P - \log Q)$ as replacement of classical divergence (here, P and Q are density operators, cf. Holevo [33] or Ohya and Petz [46]). In the setting of quantum information theory, the compensation identity is often referred to as *Donalds identity,* cf. Donalds [24]. In the classical case, the identity possibly first appeared in [54].

In Table 1 we summarize the three games discussed in this section together with an indiction of the principles this leads to when seen from the point of view of each of the two players in the game.

GTE	Player I	Player II
$H_{\max}$-game	Max Ent	Min Length
$D_{\min}$-game	Min Div	Max Saving
$I_{\max}$-game	Max Inf	Min Redundancy

Table 1. Instances of the principle of GTE

[3]This remark is due to Csiszár, oral communication at the conference "Information Theory, Cryptography and Statistics", Balatonlelle, October 2000.

5. Technical discussion of the $D_{\min}$-game

This section is intended as a relatively simple and self-contained introduction to some basic results about information projections, a theme going back to to Csiszár [14] and Čencov [10], and with numerous further publications of which we here only point to Csiszár [15]. The emphasis is on the game theoretical approach already introduced. We do not include a full discussion of the origin of the results discussed or on their further development. The interested reader should consult research cited above and the more recent references pointed to in the bibliography.

Let $(X, \mathcal{B})$ be an *abstract Borel space,* i.e. a set provided with a σ-algebra and denote by $M_+^1(X, \mathcal{B})$ the set of probability distributions on $(X, \mathcal{B})$. Let $\mathcal{P} \subseteq M_+^1(X, \mathcal{B})$ and $P_0 \in M_+^1(X, \mathcal{B})$ be given.

Consider the $D_{\min}$-game with $\mathcal{P}$ as the strategy set for Player I, with P_0 as prior distribution and with the set of all probability distributions on $(X, \mathcal{B})$ as the strategy set for Player II. Thus we choose to consider estimators or posterior distributions rather than codes as the available strategies for Player II. This will give the results a more conventional flavour which is more likely to appeal to the reader interested mainly in applications to statistics. And, of course, the concept of codes really only applies meaningfully to the discrete case. Thus, focusing on distributions enables us to present a completely general version of the $D_{\min}$-game.

As a concrete example of great interest which the reader may want to have in mind – and which we shall return to later on – we mention the case when $\mathcal{P}$ is specified by *moment conditions,* i.e. when

$$\mathcal{P} = \left\{ P \in M_+^1(X, \mathcal{B}) \mid \langle g_i, P \rangle = c_i \text{ for } i \in I \right\} \tag{19}$$

with $(g_i)_{i \in I}$ a set of measurable functions and $(c_i)_{i \in I}$ a corresponding prescribed set of scalars. The index set I is, typically, assumed to be finite.

We realize that the setting fits into the framework outlined in Section 3.1 of an information space. The knowledge base is $\mathcal{P}$, the set K of description strategies is $M_+^1(X, \mathcal{B})$ and the objective function is the map $(P, R) \curvearrowright D(P \,\|\, P_0 \rightsquigarrow R)$. As this function may assume negative values, the proper setting is that of a *relative* information space which we may identify by the pair $(\mathcal{P}, P_0)$. Distributions in $\mathcal{P}$ are referred to as *consistent distributions.* A distribution Q is *essentially consistent* if there exists a sequence $(P_n)_{n \geq 1}$ of consistent distributions which converges in divergence to Q, i.e. $D(P_n \,\|\, Q) \to 0$ as $n \to \infty$.

The defining relations (8) (for divergence) and (7) combined with (4) (for estimation gain) are now taken as the basic definitions. Written explicitly, the definition we use for estimation gain is as follows:

$$D(P \parallel Q \rightsquigarrow R) = D(P \parallel Q) - D(P \parallel R). \tag{20}$$

As there are some subtleties connected with (20) (the $\infty - \infty$ problem), it is not entirely clear if all aspects of the $D_{\min}$-game make sense. However, with appropriate conventions this can in fact be achieved. We simply agree that if a suppremum is considered involving indeterminate numbers we interpret the result in the "least favourable way" as $+\infty$ and, likewise, if an infimum involves an indeterminate number, we interpret the result as giving the quantity $-\infty$. With these conventions it is clear that each of the players can assign a performance index to any specific strategy available to the player. And for Player I we can easily identify the performance index. Indeed, if Player I chooses the strategy $P \in \mathcal{P}$, the associated performance index is nothing but the divergence $D(P \parallel P_0)$ as

$$\sup_R D(P \parallel P_0 \rightsquigarrow R) = D(P \parallel P_0)$$

(here, it is understood that R ranges over all strategies available to Player II, i.e. over all probability distributions on $(X, \mathcal{B})$, and that the conventions just introduced are in force). Player I is then led to consider the quantity

$$D_{\min} = \inf_{P \in \mathcal{P}} D(P \parallel P_0)$$

and the notion of an optimal strategy for Player I makes good sense as a consistent distribution P with $D(P \parallel P_0) = D_{\min}$. If need be we stress the dependence on $(\mathcal{P}, P_0)$ by using the notation

$$D_{\min} = D(\mathcal{P} \parallel P_0). \tag{21}$$

A sequence $(P_n)_{n \geq 1}$ of consistent distributions is said to be *asymptotically optimal* if $D(P_n \parallel P_0) \to D_{\min}$ for $n \to \infty$.

As for Player II, the performance index associated with the strategy R is given by

$$\Gamma(R) = \inf_{P \in \mathcal{P}} D(P \parallel P_0 \rightsquigarrow R) \tag{22}$$

for which we also introduce the notation

$$\Gamma(R) = D(\mathcal{P} \parallel P_0 \rightsquigarrow R) \tag{23}$$

by analogy with the notation used in (21). This is the *pay-off* or, more specifically, the *estimation gain* (or even the *guaranteed estimation gain*) associated with the strategy. Player II is then led to consider the *maximal pay-off* given by

$$\Gamma_{\max} = \sup_R \Gamma(R)$$

and the notion of an optimal strategy for Player II makes good sense as a distribution R with $\Gamma(R) = \Gamma_{\max}$. Notational symmetry with (21) is obtained by introducing the notation

$$\Gamma_{\max} = \Gamma(\mathcal{P} \parallel P_0). \tag{24}$$

Clearly, $\Gamma_{\max} \le D_{\min}$. If equality holds with a finite common value, this is the *value* of the game and the game is said to be in *equilibrium.*

The first result, really Theorem 8 of [56], can be formulated in a very standard way without reference to the $D_{\min}$-game at all. For the proof we need *Pinsker's inequality* which states that

$$D(P \parallel Q) \ge \frac{1}{2} V(P,Q)^2$$

for distributions P and Q, with V denoting total variation. We also need to know that $(P,Q) \curvearrowright D(P \parallel Q)$ is jointly lower semi-continuous, even with respect to the relatively weak topology of setwise convergence. For these facts, see [58].

Theorem 1. *Let $(\mathcal{P}, P_0)$ be a relative information space and assume that $\mathcal{P}$ is convex and that $D_{\min} < \infty$. Then there exists a unique distribution Q such that, for every $P \in \mathcal{P}$, the Pythagorean inequality holds, i.e.*

$$D(P \parallel P_0) \ge D(P \parallel Q) + D_{\min}. \tag{25}$$

Proof. Let $(P_n)_{n\ge1}$ be asymptotically optimal. By the compensation identity, we find that for every $n \ge 1$, $m \ge 1$, and with $P_{n,m}$ denoting the distribution $\frac{1}{2}P_n + \frac{1}{2}P_m$,

$$\begin{aligned} &D_{\min} + \frac{1}{2}D(P_n \parallel P_{n,m}) + \frac{1}{2}D(P_m \parallel P_{n,m}) \\ &\le D(P_{n,m} \parallel P_0) + \frac{1}{2}D(P_n \parallel P_{n,m}) + \frac{1}{2}D(P_m \parallel P_{n,m}) \\ &= \frac{1}{2}D(P_n \parallel P_0) + \frac{1}{2}D(P_m \parallel P_0). \end{aligned}$$

As this last quantity converges to $D_{\min}$, we conclude by Pinskers inequality – which shows that both $D(P_n \parallel P_{n,m})$ and $D(P_m \parallel P_{n,m})$ are lower bounded by $\frac{1}{8}V^2(P_n, P_m)$ – that $(P_n)_{n\geq 1}$ is a Cauchy-sequence w.r.t. total variation. By completeness of total variation, there exists a probability distribution Q such that $P_n \to Q$ in total variation, i.e. $V(P_n, Q) \to 0$. By a standard argument involving the mixing of two sequences, we realize that Q is independent of the particular asymptotically optimal sequence.

Now choose a sequence $(P_n)_{n\geq 1} \subseteq \mathcal{P}$ which "converges fast" in the sense that

$$n\big(D(P_n \parallel P_0) - D_{\min}\big) \to 0.$$

We shall use this auxiliary sequence to establish (25). To do so, fix $P \in \mathcal{P}$ and consider the distributions $Q_n, n \geq 1$ given by

$$Q_n = \left(1 - \frac{1}{n}\right) P_n + \frac{1}{n} P.$$

Clearly, $D(Q_n \parallel P_0) \geq D_{\min}$. Again appealing to the compensation identity, we find that

$$\begin{aligned} D_{\min} + \frac{1}{n} D(P \parallel Q_n) &\leq D(Q_n \parallel P_0) + \left(1 - \frac{1}{n}\right) D(P_n \parallel Q_n) + \frac{1}{n} D(P \parallel Q_n) \\ &= \left(1 - \frac{1}{n}\right) D(P_n \parallel P_0) + \frac{1}{n} D(P \parallel P_0) \end{aligned}$$

and therefore,

$$D(P \parallel P_0) + n\big(D(P_n \parallel P_0) - D_{\min}\big) \geq D(P \parallel Q_n) + D(P_n \parallel P_0).$$

Now then, (25) follows from the "fast convergence" of (P_n) and from the lower semi-continuity of $R \curvearrowright D(P \parallel R)$ (as Q_n converges setwise, even in total variation, to Q).

To finish the proof we note that the Pythagorean inequality uniquely characterizes Q. Indeed, if $(P_n)_{n\geq 1}$ is asymptotically optimal, then P_n converges in divergence to Q. In particular, P_n converges in total variation to Q. ■

Note the somewhat peculiar aspect of the above proof: The first part follows the argument in the proof of Csiszár, [14, Theorem 2.1] and establishes as an auxiliary result a not-so-strong form for convergence of an asymptotically optimal sequence and then, this is strengthened in the last part of the proof to a much stronger, and information theoretically more appropriate form of convergence. The fact that convergence in divergence (for ordinary sequences as here considered, although not for nets) is a topological notion of convergence follows from a well known fact of general topology (Kisynski's theorem, cf. [42]) but a more direct and intrinsicly information theoretical analysis of this important observation is possible as demonstrated by Harremoës, cf. [28], [29].

Following Csiszár [15], we say that Q of Theorem 1 is the *generalized I-projection* (of P_0 on $\mathcal{P}$). We use the notation $P_0(\cdot \mid \mathcal{P})$ for this distribution. The notation is justified by the fact that the concept can be viewed as a generalization of ordinary conditioning as we shall comment on later in more detail. If Q happens to be consistent, this is the (standard) *I-projection* (of P_0 on $\mathcal{P}$). A distribution Q is called the *relative centre of attraction* (of $\mathcal{P}$ w.r.t. P_0) if $D(P_n \parallel Q) \to 0$ for every asymptotically optimal sequence $(P_n)_{n\geq 1}$.[4]

Let us collect some immediate consequences of Theorem 1. For some of these we find it convenient to introduce the *scope* of a distribution. Definition and notation is given by:

$$\operatorname{scope}(Q) = \big\{ P \mid D(P \parallel Q) < \infty \big\}.$$

Corollary 1. *With assumptions as in Theorem 1,* $Q = P_0(\cdot \mid \mathcal{P})$ *can also be characterized as the relative centre of attraction. Furthermore,* Q *is essentially consistent and* $D(Q \parallel P_0) \leq D_{\min}$*, in particular,* $Q \in \operatorname{scope}(P_0)$.

Proof. The characterization of Q as the relative centre of attraction follows from (25) as also noted in the final part of the proof of Theorem 1. The remaining parts follow from lower semi-continuity of divergence and, again, from (25). ∎

We point out that $\mathcal{P} \cap \operatorname{scope}(Q)$ may be strictly larger than $\mathcal{P} \cap \operatorname{scope}(P_0)$. This is easily seen by exploiting well known examples going back to Csiszár [12] which show that "transitivity" may fail (i.e., we cannot conclude $P_1 \in \operatorname{scope}(P_3)$ from $P_1 \in \operatorname{scope}(P_2)$ and $P_2 \in \operatorname{scope}(P_3)$).

Simple examples show that the notion of relative centre of attraction is more general than the notion of generalized I-projection. However, the

[4]The analogous "absolute" notion refers to entropy, cf. [59]. These notions were introduced in [56] except for a weaker requirement of convergence than convergence in divergence.

two concepts are equivalent under very mild and natural conditions as we have seen in Theorem 1. Therefore, one will normally only need to refer to the well established and geometrically appealing notion of projection. An important problem then is to determine the generalized I-projection in specific cases of interest. As we shall see, the game theoretical approach is helpful in this respect.

Often, we need a stronger finiteness condition than the one appearing in Theorem 1. We say that $(\mathcal{P}, P_0)$ is *D-finite* if $D(P \parallel P_0) < \infty$ for every $P \in \mathcal{P}$, i.e. if $\mathcal{P} \subseteq \text{scope}\,(P_0)$. Associated with any model $(\mathcal{P}, P_0)$ we may consider the restriction to the scope of P_0, i.e. the model $(\mathcal{P}_0, P_0)$ with

$$\mathcal{P}_0 = \mathcal{P} \cap \text{scope}\,(P_0).$$

Clearly, this model is D-finite.

The proof of the following simple result (adapted from [15, Lemma 3.2]) is left to the reader:

Proposition 1. *Let $(\mathcal{P}, P_0)$ be given with $\mathcal{P}$ convex and $D_{\min}(\mathcal{P} \parallel P_0) < \infty$ and denote by $\mathcal{P}_0$ the restriction $\mathcal{P} \cap \text{scope}\,(P_0)$.*

Then $\mathcal{P}_0$ is convex, non-empty and $D_{\min}(\mathcal{P}_0 \parallel P_0) = D_{\min}(\mathcal{P} \parallel P_0)$. Furthermore, $(\mathcal{P}, P_0)$ and $(\mathcal{P}_0, P_0)$ have the same generalized I-projection.

Under the condition of D-finiteness we can strengthen the result of Theorem 1 as follows (essential parts are from Theorem 9 of [56]):

Theorem 2. *Let $(\mathcal{P}, P_0)$ be a D-finite relative information space with $\mathcal{P}$ convex. Then the $D_{\min}$-game is in equilibrium and Q, defined as the generalized I-projection of P_0 on $\mathcal{P}$, can also be characterized as the unique optimal estimator for Player II. Furthermore, for every distribution R the inequality*

$$\Gamma(R) \le \Gamma_{\max} - D(Q \parallel R) \tag{26}$$

holds. Finally, if $R \in \mathcal{P}$, then

$$V(Q, R) \le \sqrt{D(R \parallel P_0) - \Gamma(R)}. \tag{27}$$

Proof. We may write the Pythagorean inequality in the form $D(P \parallel P_0 \rightsquigarrow Q) \ge D_{\min}$ as $\mathcal{P} \subseteq \text{scope}\,(P_0)$. Thus, by (25), $\Gamma(Q) \ge D_{\min}$, and as the reverse inequality is trivial, the game is in equilibrium and Q is an optimal estimator. Let $(P_n)_{n\geq 1}$ be an asymptotically optimal sequence. Consider a distribution R with $\Gamma(R) > -\infty$ (otherwise, (26) is trivial). Then,

$$\begin{aligned}\Gamma(R) &\leq \limsup_{n\to\infty} D(P_n \parallel P_0 \rightsquigarrow R) \\ &= \limsup_{n\to\infty} \big(D(P_n \parallel P_0) - D(P_n \parallel R)\big) \\ &= D_{\min} - \liminf_{n\to\infty} D(P_n \parallel R) \\ &\leq D_{\min} - D(Q \parallel R)\end{aligned}$$

by lower semi-continuity and (26) follows.

The uniqueness of Q viewed as an optimal estimator follows from (26).

Combining (25) and (26) and using Pinsker's inequality, both for $D(R \parallel Q)$ and for $D(Q \parallel R)$, (27) follows. ■

The equilibrium distribution of Theorem 2 can often be found by exploiting a general equilibrium concept of game theory. Consider the D-min game associated with $(\mathcal{P}, P_0)$ and assume to begin with only that $D_{\min} < \infty$. Following normal conventions of game theory – here only for two-person games – a pair of strategies for the players is a *Nash equilibrium pair* if, fixing the one strategy, the other player cannot gain anything by changing his strategy.

For our situation, the requirements on a pair (P^*, R^*) with $P^* \in \mathcal{P}$ and $R^* \in M_+^1(X, \mathcal{B})$ is that the *saddle value inequalities*

$$D(P^* \parallel P_0 \rightsquigarrow R) \leq D(P^* \parallel P_0 \rightsquigarrow R^*) \text{ for } R \in M_+^1(X, \mathcal{B}) \tag{28}$$

and

$$D(P \parallel P_0 \rightsquigarrow R^*) \geq D(P^* \parallel P_0 \rightsquigarrow R^*) \text{ for } P \in \mathcal{P} \tag{29}$$

hold. By (29) applied to a $P \in \mathcal{P}$ with $D(P \parallel P_0) < \infty$ we see that $D(P^* \parallel P_0) < \infty$. Then, from (28), it follows that $D(P^* \parallel P_0 \rightsquigarrow P^*) \leq D(P^* \parallel P_0 \rightsquigarrow R^*)$, i.e. $D(P^* \parallel P_0) \leq D(P^* \parallel P_0) - D(P^* \parallel R^*)$ and we conclude that $P^* = R^*$ must hold. But then (28) is automatic (whatever P^* is) and the requirement (29) becomes

$$D(P \parallel P_0 \rightsquigarrow P^*) \geq D(P^* \parallel P_0) \text{ for } P \in \mathcal{P}. \tag{30}$$

Based on this analysis we agree to call P^* a *Nash equilibrium distribution* (having Player I in mind) or a *Nash equilibrium estimator* (having Player II in mind) if $P^* \in \mathcal{P}$, if $\mathcal{P} \subseteq \text{scope}\,(P_0)$ and if (30) holds. The condition $\mathcal{P} \subseteq \text{scope}\,(P_0)$ is a finiteness condition which ensures that (30) is meaningful. As the overall assumption $D_{\min} < \infty$ is in force, we realize that $P^* \in \text{scope}\,(P_0)$ for a Nash equilibrium distribution.

A Nash equilibrium distribution may not exist, but we realize that if it does, there is only one such distribution so that we may talk of *the* Nash equilibrium distribution in such cases. Indeed, if $P^* \in \mathcal{P}$ and $P^{**} \in \mathcal{P}$ are both Nash equilibrium distributions, it follows from (30) that

$$D(P^{**} \parallel P_0) - D(P^{**} \parallel P^*) \geq D(P^* \parallel P_0)$$

and that

$$D(P^* \parallel P_0) - D(P^* \parallel P^{**}) \geq D(P^{**} \parallel P_0)$$

hold. Clearly, this leads to the inequality $0 \geq D(P^* \parallel P^{**}) + D(P^{**} \parallel P^*)$, hence $P^* = P^{**}$.

Let us investigate what can be said when the Nash equilibrium distribution exists:

Theorem 3. *Let $\mathcal{P}$ be any set of distributions and P_0 any prior distribution such that $D_{\min} < \infty$. Assume that the Nash equilibrium distribution, Q, exists. Then the $D_{\min}$-game for $(\mathcal{P}, P_0)$ is in equilibrium and Q is the unique optimal strategy for Player I as well as the unique optimal strategy for Player II. In other words, Q has minimum divergence to P_0 among all consistent distributions and, considered as an estimator, Q achieves the maximal estimation gain. Furthermore, the inequalities of Theorems 1 and 2 hold.*

Before the proof we point out that $\mathcal{P}$ is not assumed to be convex for this result. The distribution Q in Theorem 3 is the *I-projection of P_0 on $\mathcal{P}$* (the qualifying "generalized" is superfluous as Q in the above result is consistent by assumption).

Proof. As $\Gamma(Q) \leq D(Q \parallel P_0)$ holds generally, we conclude from the inequality $\Gamma(Q) \geq D(Q \parallel P_0)$ that $\Gamma(Q) = D(Q \parallel P_0) = \Gamma_{\max} = D_{\min}$. The inequality (25) follows as $\Gamma(Q) \geq D_{\min}$.

And (26) follows as, for any estimator R with $D(Q \parallel R) < \infty$,

$$\Gamma(R) + D(Q \parallel R) \leq D(Q \parallel P_0 \rightsquigarrow R) + D(Q \parallel R) = D(Q \parallel P_0) = D_{\min}.$$

(Note that for the first equality sign we need the assumption $D(Q \parallel R) < \infty$).

The uniqueness assertions now follow from (25) and (26). ■

One feature of this result is that there is no *divergence loss*. By this we mean that $D(Q \parallel P_0) = D_{\min}$ must hold for the situation covered by Theorem 3, whereas, for the situation of Theorems 1 or 2 there is a possibility that $D(Q \parallel P_0) \leq D_{\min}$ holds with strict inequality.

The inequality (30) may often be fulfilled in a trivial way. What we have in mind is cases with a very robust estimator Q in the sense that the estimation gain $D(P \parallel P_0 \rightsquigarrow Q)$ is independent of which distribution is chosen by Player I. To be precise, the distribution Q is *pay-off stable* if $D(P \parallel P_0 \rightsquigarrow Q)$ is well defined and finite for all $P \in \mathcal{P}$ and, furthermore, independent of $P \in \mathcal{P}$. So we demand that $D(P \parallel P_0)$ and $D(P \parallel Q)$ are finite for all $P \in \mathcal{P}$ and that, for some constant π, $D(P \parallel P_0 \rightsquigarrow Q) = \pi$ for all $P \in \mathcal{P}$. We shall also say that Q is pay-off stable at the *level* π.

Note that, in contrast to what was the case when we looked at the notion of Nash equilibrium, we do not require that a pay-off stable distribution Q is consistent. But if it is, we are in a special, desirable situation:

Theorem 4. *Let $\mathcal{P}$ and P_0 be given and assume that Q is a consistent pay-off-stable distribution. Then Q is the Nash equilibrium estimator, hence also the I-projection of P_0 on $\mathcal{P}$. Furthermore, the Pythagorean inequality* (25) *holds with equality for every consistent distribution P.*

We leave the simple proof to the reader.

Theorem 4 (and Proposition 1) point to a strategy which often works in the search for the I-projection: First identify pay-off stable distributions associated with $(\mathcal{P}, P_0)$ (or with the corresponding restriction to scope(P_0)) and then select a consistent distribution among these. If you succeed with this, the desired I-projection has been found. The search for pay-off stable distributions is facilitated by the lemma below which takes into account that often care has to be taken when working with Radon–Nikodym derivatives in order to express estimation gain in an adequate form.

Lemma 1. *Let P, Q and P_0 be distributions. Assume that $Q \ll P_0$ and that there exists a version ϕ of the Radon–Nikodym derivative $\frac{dQ}{dP_0}$ such that $\phi > 0$ a.e. $[P]$. Assume further that $D(P \parallel P_0) < \infty$. Then $\int \log \phi \, dP = \langle \log \phi, P \rangle$ is a well-defined number in $[-\infty, \infty[$ and*

$$\langle \log \phi, P \rangle = D(P \parallel P_0 \rightsquigarrow Q). \tag{31}$$

Proof. First we prove that $P \ll Q$. To see this, assume that $Q(A) = 0$ for a set $A \in \mathcal{B}$. Then $\int_A \phi \, dP_0 = 0$, hence $P_0\big(A \cap \{\phi > 0\}\big) = 0$ and as

$P \ll P_0$, also $P\big(A \cap \{\phi > 0\}\big) = 0$ holds. As $P\big(\{\phi = 0\}\big) = 0$, we conclude as desired that $P(A) = 0$.

Let f be any version of the Radon–Nikodym derivative $\frac{dP}{dP_0}$ and let g be any version of the Radon–Nikodym derivative $\frac{dP}{dQ}$. As $P\big(\{g = 0\}\big) = 0$, we find that

$$\begin{aligned} D(P \parallel P_0) - D(P \parallel Q) &= \int \log f \, dP - \int \log g \, dP \\ &= \int_{\{g>0\}} \log f \, dP - \int_{\{g>0\}} \log g \, dP \\ &= \int_{\{g>0\}} \log \frac{f}{g} \, dP = \int_{\{g>0\}} \log \frac{g\phi}{g} dP \\ &= \int_{\{g>0\}} \log \phi \, dP = \int \log \phi \, dP \end{aligned}$$

which leads to the result. ■

We now aim at discussing in more detail sets of distributions specified by moment constraints. It is well known that the classical *exponential families* are of significance in this respect. Here we suggest to associate an "exponential family" with any relative information space. This follows a suggestion hinted at in [56]. The precise definition is as follows: The *exponential family* associated with the relative information space $(\mathcal{P}, P_0)$ is the set of distributions Q with $Q \ll P_0$ for which there exists a finite constant π and a version ϕ of the Radon–Nikodym derivative $\frac{dQ}{dP_0}$ such that $\langle \log \phi, P \rangle = \pi$ for every $P \in \mathcal{P}$. Expressed briefly,

$$\mathcal{E} = \left\{ Q \ll P_0 \,\middle|\, \exists \pi \in \mathbb{R} \; \exists \phi = \frac{dQ}{dP_0} \; \forall P \in \mathcal{P} : \langle \log \phi, P \rangle = \pi \right\} \tag{32}$$

Note that if $Q \in \mathcal{E}(\mathcal{P}, P_0)$ then every $P \in \mathcal{P}$ is concentrated on the set $\{\phi > 0\}$. Also note that the exponential family is non-empty as it always contains the distribution P_0.

Theorem 5. *Let $(\mathcal{P}, P_0)$ be given with $\mathcal{P}$ convex and $D_{\min} < \infty$. Then the exponential family $\mathcal{E}(\mathcal{P}, P_0)$ contains at most one consistent distribution and this must be the I-projection of P_0 on $\mathcal{P}$.*

Proof. Put $\mathcal{P}_0 = \mathcal{P} \cap \text{scope}\,(P_0)$. Assume that $Q \in \mathcal{P} \cap \mathcal{E}(\mathcal{P}, P_0)$ and let π and ϕ be as in (32). In particular, $\langle \log \phi, Q \rangle = \pi$, i.e. $D(Q \parallel P_0) = \pi$, hence $Q \in \text{scope}\,(P_0)$ and $Q \in \mathcal{P}_0$ follows. By Lemma 1 we realize that Q is pay-off stable for $\mathcal{P}_0$. Then, as $Q \in \mathcal{P}_0$, Theorem 4 applies and we conclude that $Q = P_0(\cdot \mid \mathcal{P}_0)$. Finally, from Proposition 1, $Q = P_0(\cdot \mid \mathcal{P})$ follows. ∎

Consider a knowledge base $\mathcal{P}$ given by moment constraints (19) and let P_0 be a prior distribution. Clearly, if $\log \frac{dQ}{dP_0}$ is a linear combination of the constant function 1 and the functions g_i, then $Q \in \mathcal{E}(\mathcal{P}, P_0)$. Therefore, if there exist constants β_i such that the quantity Z given by

$$Z = \int_X \exp\left(\sum_{i \in I} \beta_i g_i\right) dP_0 \tag{33}$$

is finite, then Q defined by

$$Q(A) = \frac{1}{Z} \int_A \exp\left(\sum_{i \in I} \beta_i g_i\right) dP_0 \tag{34}$$

is a distribution in the exponential family $\mathcal{E}(\mathcal{P}, P_0)$. If the constants β_i can be adapted so that the resulting distribution Q is consistent, Theorem 5 applies and we have found the I-projection $P_0(\cdot \mid \mathcal{P})$. "Normally" – but not always – this can be achieved. Many examples are contained in Kapur [39]. Note that the approach adopted here is direct and quite expedient – once the appropriate theory has been developed. In contrast to several other authors, the approach does not exploit general optimization techniques (Lagrange multipliers etc.) but relies on intrinsic information theoretical considerations.

A thorough recent treatment of problems related to the general case with $\mathcal{P}$ specified by moment constraints is given by Csiszár and Matúš, cf. [18] and [19]. These authors introduce an *extended exponential family.* This corresponds to a consideration of the union of those families one obtains from (32) by considering all sets $\mathcal{P}$ which may be obtained by varying the set of constants (c_i) defining $\mathcal{P}$, cf. (19).

We end by considering the special case with

$$\mathcal{P} = \left\{ P \in M_+^1(X, \mathcal{B}) \mid P(A) = 1 \right\}.$$

for $A \in \mathcal{B}$ and $P_0 \in M_+^1(X, \mathcal{B})$ with $0 < P_0(A) < 1$. This corresponds to a knowledge base given by just one moment condition: $\mathcal{P} = \{ P \mid$

$\langle 1_A, P\rangle = 1\}$ with 1_A the indicator function of A. Following the approach above we readily see that any mixture of the two conditional distributions $P_0(\cdot \mid A)$ and $P_0(\cdot \mid A^c)$ belongs to the corresponding exponential family (here A^c denotes the complement of A). Theorem 5 then tells us that $P_0(\cdot \mid \mathcal{P}) = P_0(\cdot \mid A)$. Thus, in a very simple way, the concept of I-projection may be conceived as a generalization of the classical concept of conditioning.

The example illustrates a somewhat peculiar aspect. Put $Q = P_0(\cdot \mid A)$ and assume that there exists $x_0 \in A$ such that $P(x_0) = 0$. Denote by δ_0 a unit mass at x_0. Then, by our conventions, $\Gamma(Q) = -\infty$ as the infimum in (22) contains the indeterminate number $D(\delta_0 \parallel P_0 \rightsquigarrow Q) = \infty - \infty$. Thus Q is not an optimal strategy for Player II for $(\mathcal{P}, P_0)$ whereas Q is optimal for $\big(\mathcal{P} \cap \operatorname{scope}(P_0), P_0\big)$. This somewhat undesirable situation is connected with the choice of the defining relation (20). A more subtle analysis will, instead, work with the quantity $\langle \log \phi, P\rangle$ which also appeared above, cf. (32). However, one then has to take the possibility of the various versions of $\phi = \frac{dQ}{dP_0}$ into account. Such a more refined analysis is taken up in Harremoës [29].

Acknowledgements. The author has had useful discussions on technical questions and issues of terminology with Peter Harremoës and has received useful criticism from one of the editors, Imre Csiszár, which improved the exposition and the precision of the presentation.

References

[1] S. I. Amari, Information geometry on hierarchy of probability distributions, *IEEE Trans. Inform. Theory,* **47** (2001), 1701–1711.

[2] D. Applebaum, *Probability and Information. An integrated approach,* Cambridge Univ. Press (Cambridge, 1996).

[3] J. P. Aubin, *Optima and equilibria. An introduction to nonlinear analysis,* Springer (Berlin, 1993).

[4] A. R. Barron, Entropy and the central limit theorem, *Ann. Probab.,* **14(1)** (1986), 336–342.

[5] A. R. Barron and O. Johnson, Fisher information inequalities and the central limit theorem, submitted for publication, *Probab. Theory Relat. Fields,* **129** (2004), 391–409.

[6] F. Bellini and M. Frittelli, On the existence of minimax martingale measures, *Mathematical Finance*, **12** (2002), 1–21.

[7] L. M. Bregman, The relaxation method of finding the common point of convex sets and its application to the solution of problems in convex programming, *USSR Comput. Math. and Math. Phys.*, **7** (1967), 200–217. Translated from Russian.

[8] M. Broom, Using game theory to model the evolution of information: An illustrative game, *Entropy,* **4** (2002), 35–46. Online at http://www.unibas.ch/mdpi/entropy/.

[9] N. N. Čencov, A nonsymmetric distance between probability distributions, entropy and the Pythagorean theorem. *Math. Zametki,* **4** (1968), 323–332 (in Russian).

[10] N. N. Čencov, *Statistical decision rules and optimal inference,* Nauka (Moscow, 1972), in Russian, translation in "Translations of Mathematical Monographs", 53. American Mathematical Society (1982).

[11] T. M. Cover and J. A. Thomas, *Elements of Information Theory,* Wiley (New York, 1991).

[12] I. Csiszár, Informationstheoretische Konvergenzbegriffe im Raum der Wahrscheinlichkeitverteilung, *A Magyar Tudományos Akadémia Matematikai Kutató Intézetének Közleményei,* **7** (1962), 137–158.

[13] I. Csiszár, A class of measures of informativity of observation channels, *Period. Math. Hungar.,* **2** (1972), 191–213.

[14] I. Csiszár, I-divergence geometry of probability distributions and minimization problems, *Ann. Probab.,* **3** (1975), 146–158.

[15] I. Csiszár, Sanov property, generalized I-projection and a conditional limit theorem, *Ann. Probab.,* **12** (1984), 768–793.

[16] I. Csiszár, Why least squares and maximum entropy? an axiomatic approach to inference for linear inverse problems, *Ann. Stat.,* **19** (1991), 2032–2066.

[17] I. Csiszár and J. Körner, *Information Theory: Coding Theorems for Discrete Memoryless Systems,* Academic (New York, 1981).

[18] I. Csiszár and F. Matúš, Convex cores of measures on $\mathbb{R}^d$, *Stud. Sci. Math. Hungar.,* **38** (2001), 177–190.

[19] I. Csiszár and F. Matúš, Information projections revisited, *IEEE Trans. Inform. Theory,* **49** (2003), 1474–1490.

[20] L. D. Davisson and A. Leon-Garcia, A source matching approach to finding minimax codes, *IEEE Trans. Inform. Theory,* **26** (1980), 166–174.

[21] F. Delbaen, P. Grandits, T. Rheinlaender, D. Samperi, M. Schweizer and C. Stricker, Exponential hedging and entropic penalties, *Mathematical Finance,* **12** (2002), 99–123.

[22] B. de Finetti, *Theory of Probability,* Wiley (London, 1974). Italian original 1970.

[23] A. Dembo and O. Zeitouni, *Large Deviations Techniques and Applications.* Jones and Bartlett Publishers International, Boston, 1993.

[24] M. J. Donald, On the relative entropy. *Commun. Math. Phys.,* **105** (1985), 13–34.

[25] Z. Drezner and H. Hamacher, editors, *Facility location. Applications and Theory,* Springer (Berlin, 2002).

[26] P. D. Grünwald and A. P. Dawid, Game theory, maximum entropy, minimum discrepancy, and robust bayesian decision theory, *Ann. Stat.,* **32** (2004), 1367–1433.

[27] P. Harremoës, Binomial and Poisson distributions as maximum entropy distributions. *IEEE Trans. Inform. Theory,* **47(5)** (July 2001), 2039–2041.

[28] P. Harremoës, The Information Topology, in: *Proceedings IEEE International Symposium on Information Theory,* IEEE (2002), p. 431.

[29] P. Harremoës, *Information Topologies with Applications* (2003), in this volume, pp. 113–150.

[30] P. Harremoës and F. Topsøe, Unified approach to optimization techniques in Shannon theory, in: *Proceedings, 2002 IEEE International Symposium on Information Theory,* IEEE (2002), p. 238.

[31] P. Harremoës and F. Topsøe, Maximum entropy fundamentals. *Entropy,* **3** (Sept. 2001), 191–226, http://www.unibas.ch/mdpi/entropy/ [online].

[32] D. Haussler, A general minimax result for relative entropy, *IEEE Trans. Inform. Theory,* **43** (1997), 1276–1280.

[33] A. S. Holevo, *Statistical Structure of Quantum Theory,* Springer (Berlin, 2001).

[34] E. T. Jaynes, Webpage maintained by L. Brethorst, dedicated to Jaynes work, available online from http://bayes.wustl.edu.

[35] E. T. Jaynes, Information theory and statistical mechanics, I and II, *Physical Reviews,* **106** and **108** (1957), 620–630 and 171–190.

[36] E. T. Jaynes, Clearing up mysteries – the original goal, in: J Skilling, editor, *Maximum Entropy and Bayesian Methods,* Kluwer (Dordrecht, 1989).

[37] E. T. Jaynes, *Probability Theory – The Logic of Science,* Cambridge University Press (Cambridge, 2003).

[38] A. Jessop, *Informed Assessments, an Introduction to Information, Entropy and Statistics,* Ellis Horwood (New York, 1995).

[39] J. N. Kapur, *Maximum Entropy Models in Science and Engineering,* Wiley (New York, 1993), first edition 1989.

[40] D. Kazakos, Robust noiceless source coding through a game theoretic approach, *IEEE Trans. Inform. Theory,* **29** (1983), 577–583.

[41] J. L. Kelly, A new interpretation of information rate, *Bell System Technical Journal,* **35** (1956), 917–926.

[42] J. Kisynski, Convergence du typè 1, *Colloq. Math.,* **7** (1960), 205–211.

[43] S. Kullback, *Informaton Theory and Statistics,* Wiley (New York, 1959).

[44] S. Kullback and R. Leibler, On information and sufficiency, *Ann. Math. Statist.,* **22** (1951), 79–86.

[45] Yu. V. Linnik, An information-theoretic proof of the central limit theorem with Lindeberg condition, *Theory Probab. Appl.*, **4** (1959), 288–299.

[46] M. Ohya and D. Petz, *Quantum Entropy and Its Use,* Springer (Berlin, Heidelberg, New York, 1993).

[47] E. Pfaffelhuber, Minimax information gain and minimum discrimination principle, in: I. Csiszár and P. Elias, editors, *Topics in Information Theory,* volume 16 of *Colloquia Mathematica Societatis János Bolyai,* János Bolyai Mathematical Society and North-Holland (1977), pp. 493–519.

[48] J. Rissanen, A. Barron and B. Yu, The minimum description length principle in coding and modeling, *IEEE Trans. Inform. Theory,* **44** (1998), 2743–2760.

[49] B. Ya. Ryabko, Comments on "a source matching approach to finding minimax codes", *IEEE Trans. Inform. Theory,* **27** (1981), 780–781. Including also the ensuing Editor's Note.

[50] G. Shafer and V. Vovk, *Probability and finance. It's only a game!* Wiley (Chichester, 2001).

[51] C. E. Shannon, A mathematical theory of communication. *Bell Syst. Tech. J.,* **27** (1948), 379–423 and 623–656.

[52] P. D. Straffin, *Game Theory and Strategy,* volume 36 of *New Mathematical Libary.* Mathematical Ass. of America, 1993.

[53] J. J. Sylvester, A question in the geometry of situation, *Quarterly Journal of Pure and Applied Mathematics,* **1** (1857), 79.

[54] F. Topsøe, An information theoretical identity and a problem involving capacity. *Studia Scientiarum Mathematicarum Hungarica,* **2** (1967), 291–292.

[55] F. Topsøe, A new proof of a result concerning computation of the capacity for a discrete channel, *Z. Wahrscheinlichkeitstheorie verw. Geb.,* **22** (1972), 166–168.

[56] F. Topsøe, Information theoretical optimization techniques. *Kybernetika,* **15** (1979), 8–27.

[57] F. Topsøe, Game theoretical equilibrium, maximum entropy and minimum information discrimination, in: A. Mohammad-Djafari and G. Demoments, editors, *Maximum Entropy and Bayesian Methods,* Kluwer Academic Publishers (Dordrecht, Boston, London, 1993), pp. 15–23.

[58] F. Topsøe, Basic concepts, identities and inequalities – the toolkit of information theory, *Entropy,* **3** (2001), 162–190. http://www.unibas.ch/mdpi/entropy/ [online].

[59] F. Topsøe, Maximum entropy versus minimum risk and applications to some classical discrete distributions, *IEEE Trans. Inform. Theory,* **48** (2002), 2368–2376.

[60] J. von Neumann and O. Morgenstern, *Theory of Games and Economic Behavior,* Princeton University Press (Princeton, 1947), 2nd. edition.

Flemming Topsøe

Department of Mathematics
University of Copenhagen

topsoe@math.ku.dk

BOLYAI SOCIETY
MATHEMATICAL STUDIES, 16

Entropy, Search, Complexity, pp. 209–232.

Analysis of Sorting Algorithms by Kolmogorov Complexity (A Survey)

PAUL VITÁNYI*

Recently, many results on the computational complexity of sorting algorithms were obtained using Kolmogorov complexity (the incompressibility method). Especially, the usually hard average-case analysis is ammenable to this method. Here we survey such results about Bubblesort, Heapsort, Shellsort, Dobosiewicz-sort, Shakersort, and sorting with stacks and queues in sequential or parallel mode. Especially in the case of Shellsort the uses of Kolmogorov complexity surprisingly easily resolved problems that had stayed open for a long time despite strenuous attacks.

1. Introduction

We survey recent results in the analysis of sorting algorithms using a new technical tool: the incompressibility method based on Kolmogorov complexity. Complementing approaches such as the counting method and the probabilistic method, the new method is especially suited for the average-case analysis of algorithms and machine models, whereas average-case analysis is usually more difficult than worst-case analysis using the traditional methods. Obviously, the results described can be obtained using other proof methods – all true provable statements must be provable from the axioms of mathematics by the inference methods of mathematics. The question is whether a particular proof method facilitates and guides the proving effort. The following examples make clear that thinking in terms of coding and the

*Supported in part via NeuroCOLT II ESPRIT Working Group.

incompressibility method suggests simple proofs that resolve long-standing open problems. A survey of the use of the incompressibility method in combinatorics, computational complexity, and the analysis of algorithms is [16] Chapter 6, and other recent work is [2, 15].

We give some definitions to establish notation. For introduction, details, and proofs, see [16]. We write *string* to mean a finite binary string. Other finite objects can be encoded into strings in natural ways. The set of strings is denoted by $\{0,1\}^*$. Let $x, y, z \in \mathcal{N}$, where $\mathcal{N}$ denotes the set of natural numbers. Identify $\mathcal{N}$ and $\{0,1\}^*$ according to the correspondence

$$(0, \varepsilon),\ (1, 0),\ (2, 1),\ (3, 00),\ (4, 01),\ \ldots .$$

Here ε denotes the *empty word* with no letters. The *length* of x is the number of bits in the binary string x and is denoted by $l(x)$. For example, $l(010) = 3$ and $l(\varepsilon) = 0$. The emphasis is on binary sequences only for convenience; observations in every alphabet can be so encoded in a way that is 'theory neutral.'

Self-delimiting Codes: A binary string y is a *proper prefix* of a binary string x if we can write $x = yz$ for $z \neq \varepsilon$. A set $\{x, y, \ldots\} \subseteq \{0,1\}^*$ is *prefix-free* if for every pair of distinct elements in the set neither is a proper prefix of the other. A prefix-free set is also called a *prefix code.* Each binary string $x = x_1 x_2 \ldots x_n$ has a special type of prefix code, called a *self-delimiting code,*

$$\bar{x} = 1^n 0 x_1 x_2 \ldots x_n.$$

This code is self-delimiting because we can effectively determine where the code word $\bar{x}$ ends by reading it from left to right without backing up. Using this code we define the standard self-delimiting code for x to be $x' = \overline{l(x)}x$. It is easy to check that $l(\bar{x}) = 2n + 1$ and $l(x') = n + 2\log n + 1$.

Let $\langle \cdot, \cdot \rangle$ be a standard one-one mapping from $\mathcal{N} \times \mathcal{N}$ to $\mathcal{N}$, for technical reasons chosen such that $l(\langle x, y \rangle) = l(y) + l(x) + 2l(l(x)) + 1$, for example $\langle x, y \rangle = x'y = 1^{l(l(x))} 0 l(x) x y$.

Kolmogorov Complexity: Informally, the Kolmogorov complexity, or algorithmic entropy, $C(x)$ of a string x is the length (number of bits) of a shortest binary program (string) to compute x on a fixed reference universal computer (such as a particular universal Turing machine). Intuitively, $C(x)$ represents the minimal amount of information required to generate x by any effective process, [10]. The conditional Kolmogorov complexity $C(x \mid y)$ of x relative to y is defined similarly as the length of a shortest program to compute x, if y is furnished as an auxiliary input to the computation. The functions $C(\cdot)$ and $C(\cdot \mid \cdot)$, though defined in terms of a particular machine model, are machine-independent up to an additive constant (depending on the particular enumeration of Turing machines and the particular reference universal Turing machine selected). They acquire an asymptotically universal and absolute character through Church's thesis, and from the ability of universal machines to simulate one another and execute any effective process, see for example [16]. Formally:

Definition 1. Let $T_0, T_1, \ldots$ be a standard enumeration of all Turing machines. Choose a universal Turing machine U that expresses its universality in the following manner:

$$U(\langle \langle i,p \rangle, y \rangle) = T_i(\langle p, y \rangle)$$

for all i and $\langle p, y \rangle$, where p denotes a Turing program for T_i and y an auxiliary input. We fix U as our *reference universal computer* and define the *conditional Kolmogorov complexity* of x given y by

$$C(x \mid y) = \min_{q \in \{0,1\}^*} \{ l(q) : U(\langle q, y \rangle) = x \},$$

for every q (for example $q = \langle i, p \rangle$ above) and auxiliary input y. The *unconditional Kolmogorov complexity* of x is defined by $C(x) = C(x \mid \varepsilon)$. For convenience we write $C(x, y)$ for $C(\langle x, y \rangle)$, and $C(x \mid y, z)$ for $C(x \mid \langle y, z \rangle)$.

Incompressibility: First we show that the Kolmogorov complexity of a string cannot be significantly more than its length. Since there is a Turing machine, say T_i, that computes the identity function $T_i(x) \equiv x$, and by definition of universality of U we have $U(\langle i, p \rangle) = T_i(p)$. Hence, $C(x) \leq l(x) + c$ for fixed $c \leq 2 \log i + 1$ and all x. [1] [2]

It is easy to see that there are also strings that can be described by programs much shorter than themselves. For instance, the function defined by $f(1) = 2$ and $f(i) = 2^{f(i-1)}$ for $i > 1$ grows very fast, $f(k)$ is a "stack" of k twos. It is clear that for every k it is the case that $f(k)$ has complexity at most $C(k) + O(1)$.

---

[1] "$2 \log i$" and not "$\log i$" since we need to encode i in such a way that U can determine the end of the encoding. One way to do that is to use the code $1^{l(l(i))} 0 l(i) i$ which has length $2l(l(i)) + l(i) + 1 < 2 \log i$ bits.

[2] In what follows, "log" denotes the binary logarithm. "$\lfloor r \rfloor$" is the greatest integer q such that $q \leq r$.

What about incompressibility? For every n there are 2^n binary strings of lengths n, but only $\sum_{i=0}^{n-1} 2^i = 2^n - 1$ descriptions in binary string format of lengths less than n. Therefore, there is at least one binary string x of length n such that $C(x) \geq n$. We call such strings *incompressible*. The same argument holds for conditional complexity: since for every length n there are at most $2^n - 1$ binary programs of lengths $< n$, for every binary string y there is a binary string x of length n such that $C(x \mid y) \geq n$. Strings that are incompressible are patternless, since a pattern could be used to reduce the description length. Intuitively, we think of such patternless sequences as being random, and we use "random sequence" synonymously with "incompressible sequence." There is also a formal justification for this equivalence, which does not need to concern us here. Since there are few short programs, there can be only few objects of low complexity: the number of strings of length n that are compressible by at most δ bits is at least $2^n - 2^{n-\delta} + 1$.

Lemma 1. *Let δ be a positive integer. For every fixed y, every set S of cardinality m has at least $m(1 - 2^{-\delta}) + 1$ elements x with $C(x \mid y) \geq \lfloor \log m \rfloor - \delta$.*

Proof. There are $N = \sum_{i=0}^{n-1} 2^i = 2^n - 1$ binary strings of length less than n. A fortiori there are at most N elements of S that can be computed by binary programs of length less than n, given y. This implies that at least $m - N$ elements of S cannot be computed by binary programs of length less than n, given y. Substituting n by $\lfloor \log m \rfloor - \delta$ together with Definition 1 yields the lemma. ■

Lemma 2. *If A is a set, then for every y every element $x \in A$ has complexity $C(x \mid A, y) \leq \log |A| + O(1)$.*

Proof. A string $x \in A$ can be described by first describing A in $O(1)$ bits and then giving the index of x in the enumeration order of A. ■

As an example, set $S = \{x : l(x) = n\}$. Then is $|S| = 2^n$. Since $C(x) \leq n + c$ for some fixed c and all x in S, Lemma 1 demonstrates that this trivial estimate is quite sharp. If we are given S as an explicit table then we can simply enumerate its elements (in, say, lexicographical order) using a fixed program not depending on S or y. Such a fixed program can be given in $O(1)$ bits. Hence the complexity satisfies $C(x \mid S, y) \leq \log |S| + O(1)$.

Incompressibility Method: In a typical proof using the incompressibility method, one first chooses an incompressible object from the class under discussion. The argument invariably says that if a desired property does not hold, then in contrast with the assumption, the object can be compressed significantly. This yields the required contradiction. Since most objects are almost incompressible, the desired property usually also holds for almost all objects, and hence on average. Below, we demonstrate the utility of the incompressibility method to obtain simple and elegant proofs.

Average-case Complexity: For many algorithms, it is difficult to analyze the average-case complexity. Generally speaking, the difficulty comes from the fact that one has to analyze the time complexity for all inputs of a given length and then compute the average. This is a difficult task. Using the incompressibility method, we choose just one input – a representative input. Via Kolmogorov complexity, we can show that the time complexity of this input is in fact the average-case complexity of all inputs of this length. Constructing such a "representative input" is impossible, but we know it exists and this is sufficient.

In average-case analysis, the incompressibility method has an advantage over a probabilistic approach. In the latter approach, one deals with expectations or variances over some ensemble of objects. Using Kolmogorov complexity, we can reason about an incompressible individual object. Because it is incompressible it has all simple statistical properties with certainty, rather than having them hold with some (high) probability as in a probabilistic analysis. This fact greatly simplifies the resulting analysis.

2. Bubblesort

A simple introductory example of the application of the incompressibility method is the average-case analysis of Bubblesort. The classical approach can be found in [11]. It is well-known that Bubblesort uses $\Theta(n^2)$ comparisons/exchanges on the average. We present a very simple proof of this fact. The proof is based on the following intuitive idea: There are $n!$ different permutations. Given the sorting process (the insertion paths in the right order) one can recover the correct permutation from the sorted list. Hence one requires $n!$ pairwise different sorting processes. This gives a lower bound on the minimum of the maximal length of a process. We formulate the proof in the crisp format of incompressibility. In Bubblesort we make passes from left to right over the permutation to be sorted and always move the currently largest element right by exchanges between it and the right-adjacent element – if that one is smaller. We make at most $n-1$ passes, since after moving all but one element in the correct place the single remaining element must be also in its correct place (it takes two elements to be wrongly placed). The total number of exchanges is obviously at most n^2, so we only need to consider the lower bound. Let B be a Bubblesort algorithm. For a permutation π of the elements $1,\ldots,n$, we can describe the total number of exchanges by $M := \sum_{i=1}^{n-1} m_i$ where m_i is the initial distance of element $n-i$ to its final position. Note that in every pass more than one element may "bubble" right but that means simply that in the future passes of the sorting process an equal number of exchanges will be saved for the element to reach its final position. That is, every element executes a number of exchanges going right that equals precisely the initial distance between its start position to its final position. It is clear that $M \leq n^2$ for all permutations. Given $m_1,\ldots,m_{n-1}$, in that order, we can reconstruct the original permutation from the final sorted list. Since choosing a elements from a list of $b+a$ elements divides the remainder in a sequence of $a+1$ possibly empty sublists, there are

$$B(M) = \binom{M+n-2}{n-2}$$

possibilities to partition M into $n-1$ ordered non-negative summands. Therefore, we can describe π by M, n, an index of $\log B(M)$ bits to describe $m_1,\ldots,m_{n-1}$ among all partitions of M, and an program P that reconstructs π from these parameters and the final sorted list $1,\ldots,n$. Consider permutations π satisfying $C\big(\pi \mid n, B(M), P\big) \geq \log n! - \log n$. Then by Lemma 2 at least a $(1-1/n)$th fraction of all permutations of n elements have that high complexity. Under this complexity condition on π, we also have $M \geq n$. (If $M < n$ then $C\big(\pi \mid n, B(M), P\big) = O(n)$.) Since the description of π we have constructed is effective, its length must be at least $C(\pi \mid n, B, P)$. Encoding M self-delimiting, in order to be able to separate M from $B(M)$ in a concatenation of the binary descriptions, we therefore find $\log M + 2\log\log M + \log B(M) \geq n \log n - O(\log n)$. Substitute a good estimate for $\log B(M)$ (the formula used later in the Shellsort example, Section 4) divide by n, and discard the terms that vanish with n, assuming $2 < n \leq M \leq n^2$, yields $\log\big(1 + M/(n-2)\big) \geq \log n + O(1)$. By the above, this holds for at least an $(1-1/n)$th fraction of all permutations, and hence gives us an $\Omega(n^2)$ lower bound on the expected number of comparisons/exchanges.

3. Heapsort

Heapsort is a widely used sorting algorithm. One reason for its prominence is that its running time is *guaranteed* to be of order $n \log n$, and it does not require extra memory space. The method was first discovered by J. W. J. Williams, [29], and subsequently improved by R. W. Floyd [4] (see [11]). Only recently has one succeeded in giving a precise analysis of its average-case performance [23]. I. Munro has suggested a remarkably simple solution using incompressibility [18] initially reported in [16].

A "heap" can be visualized as a complete directed binary tree with possibly some rightmost nodes being removed from the deepest level. The tree has n nodes, each of which is labeled with a different key, taken from a linearly ordered domain. The largest key k_1 is at the root (on top of the heap), and each other node is labeled with a key that is less than the key of its father.

Definition 2. Let *keys* be elements of $\mathcal{N}$. An array of keys $k_1, \ldots, k_n$ is a *heap* if they are partially ordered such that

$$k_{\lfloor j/2 \rfloor} \geq k_j \quad \text{for} \quad 1 \leq \lfloor j/2 \rfloor < j \leq n.$$

Thus, $k_1 \geq k_2$, $k_1 \geq k_3$, $k_2 \geq k_4$, and so on. We consider "in place" sorting of n keys in an array $A[1 \ldots n]$ without use of additional memory.

Heapsort {Initially, $A[1 \ldots n]$ contains n keys. After sorting is completed, the keys in A will be ordered as $A[1] < A[2] < \cdots < A[n]$.}

Heapify: {Regard A as a tree: the root is in $A[1]$; the two sons of $A[i]$ are at $A[2i]$ and $A[2i+1]$, when $2i, 2i+1 \leq n$. We convert the tree in A to a heap.} **Repeat for** $i = \lfloor n/2 \rfloor, \lfloor n/2 \rfloor - 1, \ldots, 1$: {the subtree rooted at $A[i]$ is now almost a heap except for $A[i]$} push the key, say k, at $A[i]$ down the tree (determine which of the two sons of $A[i]$ possesses the greatest key, say k' in son $A[2i+j]$ with j equals 0 or 1); **if** $k' > k$ **then** put k in $A[2i+j]$ and **repeat** this process pushing k' at $A[2i+j]$ down the tree **until** the process reaches a node that does not have a son whose key is greater than the key now at the father node.

Sort: Repeat for $i = n, n-1, \ldots, 2$: {$A[1 \ldots i]$ contains the remaining heap and $A[i+1 \ldots n]$ contains the already sorted list $k_{i+1}, \ldots, k_n$ of largest elements. By definition, the element on top of the heap in $A[1]$ must be k_i.} switch the key k_i in $A[1]$ with the key k in $A[i]$, extending the sorted list to $A[i \ldots n]$. Rearrange $A[1 \ldots i-1]$ to a heap with the largest element at $A[1]$.

It is well known that the Heapify step can be performed in $O(n)$ time. It is also known that the Sort step takes no more than $O(n \log n)$ time. We analyze the precise average-case complexity of the Sort step. There are two ways of rearranging the heap: Williams's method and Floyd's method.

Williams's Method: {Initially, $A[1] = k$.}

Repeat compare the keys of k's two direct descendants; **if** m is the larger of the two **then** compare k and m; **if** $k < m$ **then** switch k and m in $A[1 \dots i-1]$ **until** $k \geq m$.

Floyd's Method: {Initially, $A[1]$ is empty.} Set $j := 1$;

while $A[j]$ is not a leaf **do**:
  if $A[2j] > A[2j+1]$ **then** $j := 2j$
  else $j := 2j + 1$;

while $k > A[j]$ **do**:
  {back up the tree until the correct position for k} $j := \lfloor j/2 \rfloor$;

move keys of $A[j]$ and each of its ancestors one node upwards;
  Set $A[j] := k$.

The difference between the two methods is as follows. Williams's method goes from the root at the top down the heap. It makes two comparisons with the son nodes and one data movement at each step until the key k reaches its final position. Floyd's method first goes from the root at the top down the heap to a leaf, making only one comparison each step. Subsequently, it goes from the bottom of the heap up the tree, making one comparison each step, until it finds the final position for key k. Then it moves the keys, shifting every ancestor of k one step up the tree. The final positions in the two methods are the same; therefore both algorithms make the same number of key movements. Note that in the last step of Floyd's algorithm, one needs to move the keys carefully upward the tree, avoiding swaps that would double the number of moves.

The heap is of height $\log n$. If Williams's method uses $2d$ comparisons, then Floyd's method uses $d + 2\delta$ comparisons, where $\delta = \log n - d$. Intuitively, δ is generally very small, since most elements tend to be near the bottom of the heap. This makes it likely that Floyd's method performs better than Williams's method. We analyze whether this is the case. Assume a uniform probability distribution over the lists of n keys, so that all input lists are equally likely.

Average-case analysis in the traditional manner suffers from the problem that, starting from a uniform distribution on the lists, it is difficult to compute the distribution on the resulting initial heaps, and increasingly more difficult to compute the distributions on the sequence of decreasing-size heaps after subsequent heapsort steps. The sequence of distributions seem somehow realated, but this is hard to express and exploit in the traditional approach. In contrast, using Kolmogorov complexity we express this similarity without having to be precise about the distributions.

Theorem 1. *On average (uniform distribution), Heapsort makes $n \log n + O(n)$ data movements. Williams's method makes $2n \log n - O(n)$ comparisons on average. Floyd's method makes $n \log n + O(n)$ comparisons on average.*

Proof. Given n keys, there are $n!$ ($\approx n^n e^{-n}\sqrt{2\pi n}$ by Stirling's formula) permutations. Hence we can choose a permutation p of n keys such that

$$C(p \mid n) \geq n \log n - 2n, \tag{1}$$

justified by Theorem 1, page 212. In fact, most permutations satisfy Equation 1.

Claim 1. Let h be the heap constructed by the **Heapify** step with input p that satisfies Equation 1. Then

$$C(h \mid n) \geq n \log n - 6n. \tag{2}$$

Proof. Assume the contrary, $C(h \mid n) < n \log n - 6n$. Then we show how to describe p, using h and n, in fewer than $n \log n - 2n$ bits as follows. We will encode the **Heapify** process that constructs h from p. At each loop, when we push $k = A[i]$ down the subtree, we record the path that key k traveled: 0 indicates a left branch, 1 means a right branch, 2 means halt. In total, this requires $(n \log 3) \sum_j j/2^{j+1} \leq 2n \log 3$ bits. Given the final heap h and the above description of updating paths, we can reverse the procedure of **Heapify** and reconstruct p. Hence, $C(p \mid n) < C(h \mid n) + 2n \log 3 + O(1) < n \log n - 2n$, which is a contradiction. (The term $6n$ above can be reduced by a more careful encoding and calculation.) ■

We give a description of h using the history of the $n-1$ heap rearrangements during the Sort step. We only need to record, for $i := n-1, \ldots, 2$, at the $(n-i+1)$st round of the Sort step, the final position where $A[i]$ is inserted into the heap. Both algorithms insert $A[i]$ into the same slot using the same number of data moves, but a different number of comparisons.

We encode such a final position by describing the path from the root to the position. A path can be represented by a sequence s of 0's and 1's, with 0 indicating a left branch and 1 indicating a right branch. Each path i is encoded in self-delimiting form by giving the value $\delta_i = \log n - l(s_i)$ encoded in self-delimiting binary form, followed by the literal binary sequence s_i encoding the actual path. This description requires at most

$$l(s_i) + 2 \log \delta_i \tag{3}$$

bits. Concatenate the descriptions of all these paths into sequence H.

Claim 2. We can effectively reconstruct heap h from H and n.

Proof. Assume H is known and the fact that h is a heap on n different keys. We simulate the Sort step in reverse. Initially, $A[1 \ldots n]$ contains a sorted list with the least element in $A[1]$.

for $i := 2, \ldots, n-1$ **do:** {now $A[1 \ldots i-1]$ contains the partially constructed heap and $A[i \ldots n]$ contains the remaining sorted list with the least element in $A[i]$} Put the key of $A[i]$ into $A[1]$, while shifting every key on the $(n-i)$th path in H one position down starting from the root at $A[1]$. The last key on this path has nowhere to go and is put in the empty slot in $A[i]$.

termination {Array $A[1 \ldots n]$ contains heap h.} ■

It follows from Claim 2 that $C(h \mid n) \leq l(H) + O(1)$. Therefore, by Equation 2, we have $l(H) \geq n \log n - 6n$. By the description in Equation 3, we have

$$\sum_{i=1}^{n} (l(s_i) + 2 \log \delta_i) = \sum_{i=1}^{n} \big((\log n) - \delta_i + 2 \log \delta_i \big) \geq n \log n - 6n.$$

It follows that $\sum_{i=1}^{n} (\delta_i - 2 \log \delta_i) \leq 6n$. This is only possible if $\sum_{i=1}^{n} \delta_i = O(n)$. Therefore, the average path length is at least $\log n - c$, for some fixed constant c. In each round of the Sort step the path length equals the number of data moves. The combined total path length is at least $n \log n - nc$.

It follows that starting with heap h, Heapsort performs at least $n \log n - O(n)$ data moves. Trivially, the number of data moves is at most $n \log n$. Together this shows that Williams's method makes $2n \log n - O(n)$ key comparisons, and Floyd's method makes $n \log n + O(n)$ key comparisons.

Since most permutations are Kolmogorov random, these bounds for one random permutation p also hold for all permutations *on average*. But we can make a stronger statement. We have taken $C(p \mid n)$ at least εn below the possible maximum, for some constant $\varepsilon > 0$. Hence, a fraction of at least $1 - 2^{-\varepsilon n}$ of all permutations on n keys will satisfy the above bounds. ■

4. SHELLSORT

The question of a nontrivial general lower bound (or upper bound) on the average complexity of Shellsort (due to D. L. Shell [26]) has been open for about four decades [11, 25], and only recently such a general lower bound was obtained. The original proof using Kolmogorov complexity [12] is presented here. Later, it turned out that the argument can be translated to a counting argument [13]. It is instructive that thinking in terms of code length and Kolmogorov complexity enabled advances in this problem.

Shellsort sorts a list of n elements in p passes using a sequence of increments $h_1, \ldots, h_p$. In the kth pass the main list is divided in h_k separate sublists of length $\lceil n/h_k \rceil$, where the ith sublist consists of the elements at positions j, where $j \bmod h_k = i - 1$, of the main list ($i = 1, \ldots, h_k$). Every sublist is sorted using a straightforward insertion sort. The efficiency of the method is governed by the number of passes p and the selected increment sequence $h_1, \ldots, h_p$ with $h_p = 1$ to ensure sortedness of the final list. The original $\log n$-pass[3] increment sequence $\lfloor n/2 \rfloor, \lfloor n/4 \rfloor, \ldots, 1$ of Shell [26] uses worst case $\Theta(n^2)$ time, but Papernov and Stasevitch [19] showed that another related sequence uses $O(n^{3/2})$ and Pratt [22] extended this to a class of all nearly geometric increment sequences and proved this bound was tight. The currently best asymptotic method was found by Pratt [22]. It uses all $\log^2 n$ increments of the form $2^i 3^j < \lfloor n/2 \rfloor$ to obtain time $O(n \log^2 n)$ in the worst case. Moreover, since every pass takes at least n steps, the average complexity using Pratt's increment sequence is $\Theta(n \log^2 n)$. Incerpi and Sedgewick [5] constructed a family of increment sequences for which Shellsort runs in $O\big(n^{1+\varepsilon/\sqrt{\log n}}\big)$ time using $(8/\varepsilon^2) \log n$ passes, for every $\varepsilon > 0$. B. Chazelle (attribution in [24]) obtained the same result by generalizing Pratt's method: instead of using 2 and 3 to construct

[3] "log" denotes the binary logarithm and "ln" denotes the natural logarithm.

the increment sequence use a and $(a+1)$ to obtain a worst-case running time of $n \log^2 n (a^2/\ln^2 a)$ which is $O\big(n^{1+\varepsilon/\sqrt{\log n}}\big)$ for $\ln^2 a = O(\log n)$. Poonen [20], and Plaxton, Poonen and Suel [21], demonstrated an $\Omega\big(n^{1+\varepsilon/\sqrt{p}}\big)$ lower bound for p passes of Shellsort using any increment sequence, for some $\varepsilon > 0$; taking $p = \Omega(\log n)$ shows that the Incerpi–Sedgewick / Chazelle bounds are optimal for small p and taking p slightly larger shows a $\Theta\big(n \log^2 n/(\log\log n)^2\big)$ lower bound on the worst-case complexity of Shellsort. For the *average-case* running time Knuth [11] showed $\Theta(n^{5/3})$ for the best choice of increments in $p = 2$ passes; Yao [30] analyzed the average-case for $p = 3$ but did not obtain a simple analytic form; Yao's analysis was improved by Janson and Knuth [7] who showed $O\big(n^{23/15}\big)$ average-case running time for a particular choice of increments in $p = 3$ passes. Apart from this no nontrivial results are known for the average-case; see [11, 24, 25]. In [12, 13] a general $\Omega(pn^{1+1/p})$ lower bound was obtained on the average-case running time of p-pass Shellsort under uniform distribution of input permutations, for every $1 \le p \le n/2$.[4] This is the first advance on the problem of determining general nontrivial bounds on the *average-case* running time of Shellsort [22, 11, 30, 5, 21, 24, 25].

A Shellsort computation consists of a sequence of comparison and inversion (swapping) operations. In this analysis of the average-case lower bound we count just the total number of data movements (here inversions) executed. The same bound holds *a fortiori* for the number of comparisons.

Theorem 2. *The average number of comparisons (and also inversions for $p = o(\log n)$) in p-pass Shellsort on lists of n keys is at least $\Omega\big(pn^{1+1/p}\big)$ for every increment sequence. The average is taken with all lists of n items equally likely (uniform distribution).*

Proof. Let the list to be sorted consist of a permutation π of the elements $1, \ldots, n$. Consider a $(h_1, \ldots, h_p)$ Shellsort algorithm A where h_k is the increment in the kth pass and $h_p = 1$. For every $1 \le i \le n$ and $1 \le k \le p$, let $m_{i,k}$ be the number of elements in the h_k-increment sublist, containing element i, that are to the left of i at the beginning of pass k and are larger than i. Observe that $\sum_{i=1}^{n} m_{i,k}$ is the number of inversions in the initial permutation of pass k, and that the insertion sort in pass k requires precisely

[4] The trivial lower bound is $\Omega(pn)$ comparisons since every element needs to be compared at least once in every pass.

$\sum_{i=1}^{n}(m_{i,k}+1)$ comparisons. Let M denote the total number of inversions:

$$M := \sum_{k=1}^{p} \sum_{i=1}^{n} m_{i,k}. \tag{4}$$

Claim 3. Given all the $m_{i,k}$'s in an appropriate fixed order, we can reconstruct the original permutation π.

Proof. In general, given the $m_{i,k}$'s and the final permutation of pass k, we can reconstruct the initial permutation of pass k. ■

Let M as in (4) be a fixed number. There are $n!$ permutations of n elements. Let permutation π be an incompressible permutation having Kolmogorov complexity

$$C(\pi \mid n, A, P) \geq \log n! - \log n, \tag{5}$$

where P is the decoding program in the following discussion. There exist many such permutations by lemma 1. Clearly, there is a fixed program that on input A, P, n reconstructs π from the description of the $m_{i,k}$'s as in Claim 3. Therefore, the minimum length of the latter description, including a fixed program in $O(1)$ bits, must exceed the complexity of π:

$$C(m_{1,1}, \ldots, m_{n,p} \mid n, A, P) + O(1) \geq C(\pi \mid n, A, P). \tag{6}$$

An M as defined by (4) such that every division of M in $m_{i,k}$'s contradicts (6) would be a lower bound on the number of inversions performed. Similar to the reasoning Bubblesort example, Section 2, there are

$$D(M) := \binom{M + np - 1}{np - 1} \tag{7}$$

distinct divisions of M into np ordered nonnegative integral summands $m_{i,k}$'s. Every division can be indicated by its index j in an enumeration of these divisions. This is both obvious and an application of lemma 2. Therefore, a description of M followed by a description of j effectively describes the $m_{i,k}$'s. Fix P as the program for the reference universal machine that reconstructs the ordered list of $m_{i,k}$'s from this description. The binary length of this two-part description must by definition exceed the Kolmogorov complexity of the described object.

A minor complication is that we cannot simply concatenate two binary description parts: the result is a binary string without delimiter to indicate where one substring ends and the other one begins. Encoding the M part of the description self-delimitingly we obtain:

$$\log D(M) + \log M + 2 \log\log M + 1 \geq C(m_{1,1}, \ldots, m_{n,p} \mid n, A, P).$$

We know that $M \leq pn^2$ since every $m_{i,k} \leq n$. We can assume[5] $p < n$. Together with (5) and (6), we have

$$\log D(M) \geq \log n! - 4 \log n - 2 \log\log n - O(1). \tag{8}$$

Estimate $\log D(M)$ by [6]

$$\log \binom{M + np - 1}{np - 1} = (np - 1) \log \frac{M + np - 1}{np - 1} + M \log \frac{M + np - 1}{M}$$
$$+ \frac{1}{2} \log \frac{M + np - 1}{(np - 1)M} + O(1).$$

The second term in the right-hand side equals[7]

$$\log \left(1 + \frac{np - 1}{M}\right)^M < \log e^{np-1}$$

for all positive M and $np - 1 > 0$. Since $0 < p < n$ and $n \leq M \leq pn^2$,

$$\frac{1}{2(np-1)} \log \frac{M + np - 1}{(np - 1)M} \rightarrow 0$$

for $n \rightarrow \infty$. Therefore, $\log D(M)$ is majorized asymptotically by

$$(np - 1) \left(\log \left(\frac{M}{np - 1} + 1 \right) + \log e \right)$$

[5] Otherwise we require at least n^2 comparisons.

[6] Use the following formula ([16], p. 10),

$$\log \binom{a}{b} = b \log \frac{a}{b} + (a - b) \log \frac{a}{a - b} + \frac{1}{2} \log \frac{a}{b(a - b)} + O(1).$$

[7] Use $e^a > \left(1 + \frac{a}{b}\right)^b$ for all $a > 0$ and positive integer b.

for $n \to \infty$. Since the righthand-side of (8) is asymptotic to $n \log n$ for $n \to \infty$, this yields

$$M = \Omega(pn^{1+1/p}),$$

for $p = o(\log n)$. (More precisely, $M = \Omega(pn^{1+(1-\varepsilon)/p})$ for $p \leq (\varepsilon / \log e) \log n$ $(0 < \varepsilon < 1)$, see [13].) That is, the running time of the algorithm is as stated in the theorem for every permutation π satisfying satisfying (5). By lemma 1 at least a $(1-1/n)$-fraction of all permutations π require that high complexity. Then the following is a lower bound on the expected number of inversions of the sorting procedure:

$$\left(1 - \frac{1}{n}\right) \Omega\left(pn^{1+1/p}\right) + \frac{1}{n}\Omega(0) = \Omega\left(pn^{1+1/p}\right),$$

for $p = o(\log n)$. For $p = \Omega(\log n)$, the lower bound on the number of comparisons is trivially $pn = \Omega\left(pn^{1+1/p}\right)$. This gives us the theorem. ∎

Our lower bound on the average-case can be compared with the Plaxton–Poonen–Suel $\Omega\left(n^{1+\varepsilon/\sqrt{p}}\right)$ worst case lower bound [21]. Some special cases of the lower bound on the average-case complexity are:

1. For $p = 1$ our lower bound is asymptotically tight: it is the average number of inversions for Insertion Sort.
2. For $p = 2$, Shellsort requires $\Omega(n^{3/2})$ inversions (the tight bound is known to be $\Theta(n^{5/3})$ [11]);
3. For $p = 3$, Shellsort requires $\Omega(n^{4/3})$ inversions (the best known upper bound is $O(n^{23/15})$ in [7]);
4. For $p = \log n / \log \log n$, Shellsort requires $\Omega(n \log^2 n / \log \log n)$ inversions;
5. For $p = \log n$, Shellsort requires $\Omega(n \log n)$ comparisons on average. This is of course the lower bound of average number of comparisons for every sorting algorithm.
6. In general, for $n/2 > p = p(n) > \log n$, Shellsort requires $\Omega\left(n \cdot p(n)\right)$ comparisons (since every pass trivially makes n comparisons).

In [25] it is mentioned that the existence of an increment sequence yielding an average $O(n \log n)$ Shellsort has been open for 30 years. The above lower bound on the average shows that the number p of passes of such an increment sequence (if it exists) is precisely $p = \Theta(\log n)$; all the other possibilities are ruled out: Is there an increment sequence for $\log n$-pass Shellsort so that it runs in average-case $\Theta(n \log n)$? Can we tighten the average-case lower bound for Shellsort? The above bound is known to be not tight for $p = 2$ passes.

5. Dobosiewicz Sort and Shakersort

We look at some variants of Shellsort. Knuth [11], 1st Edition Exercise 5.2.1.40 on page 105, and Dobosiewicz [3] proposed to use only one pass of Bubblesort on each subsequence instead of sorting the subsequences at each stage. Incerpi and Sedgewick [6] used two passes of Bubblesort in each stage, one going left-to-right and the other going right-to-left. This is called Shakersort since it reminds one of shaking a cocktail. In both cases the sequence may stay unsorted, even if the last increment is 1. A final phase, a straight insertion sort, is required to guaranty a fully sorted list. Until recently, these variants have not been seriously analyzed; in [3, 6, 28, 24] mainly empirical evidence is reported, giving evidence of good running times (comparable to Shellsort) on randomly generated input key sequences of moderate length. The evidence also suggests that the worst-case running time may be quadratic. Again, let n be the number of keys to be sorted and let p be the number of passes. The $\Omega(n^{1+c/\sqrt{p}})$ worst-case lower bound of Poonen [20] holds apart from Shellsort also for the variants of it. We also have a worst-case lower bound of $\Omega(n^2)$ on Dobosiewicz sort and Shaker sort for the special case of almost geometric sequences of increments. But recently Brejova [1] proved that Shaker sort runs in $O(n^{3/2}\log^3 n)$ worst-case time for a certain sequence of increments (the first non-quadratic worst-case upper bound). Using the incompressibility method, she also proved lower bounds on the average-case running times.

Theorem 3. *There is an $\Omega(n^2/4^p)$ lower bound on the average-case running time of Shaker sort, and a $\Omega(n^2/2^p)$ lower bound on the average-case running time of Dobosiewicz sort. The avereges are taken with respect to the uniform distribution.*

Remark 1. These lower bounds (on the average-case) are better than the Poonen [20] lower bounds of $\Omega\big(n^{1+c/\sqrt{p}}\big)$ on the worst-case.

Proof. Consider Dobosiewicz sorting algorithm A (the description of A includes the number of passes p and the list of increments $h_1, \ldots, h_p$). Every comparison based sorting algorithm uses $\Omega(n \log n)$ comparisons on average. If $p > \log n - \log\log n$ then the claimed lower bound trivially holds. So we can assume that $p \leq \log n - \log\log n$. Let π be the permutation of $\{0, 1, \ldots, n-1\}$ to be sorted, and let π' be the permutation remaining after all p stages of the Dobsiewicz sort, but before the final insertion sort. If X is the number of inversions in π' then the final insertion sort takes $\Omega(X)$ time.

Claim 4. Let π be a permutation satisfying (5). Then $X = \Omega(n^2/2^p)$.

Proof. We can reconstruct π from π' given p strings of lengths n defined as follows: The jth bit of the ith string is "1" if x_j was interchanged with x_{j-h_i} in the ith phase of the algorithm (h_i is the ith increment), and "0" otherwise. Given π' and these strings in appropriate order we can simply run the p sorting phases in reverse.

Furthermore, π' can be reconstructed from its inversion table $a_0, a_2, \ldots,$ a_{n-1}, where a_i is the number of elements in list π' left of the ith position that are greater than the element in the ith position. Thus, $\sum_i a_i = X$. There are $D(X) = \binom{X+n-1}{n-1}$ ordered partitions of X into n non-negative summands. Hence, π' can be reconstructed from X and an index of $\log D(X)$ bits identifying the partition in question. Given n, we encode X self-delimiting to obtain a total description of $\log X + 2\log\log X + \log D(X)$ bits.

Therefore, with P the reconstruction program, we have shown that

$$C(\pi \mid n, A, P) \leq np + \log X + 2\log\log X + \log D(X).$$

Estimating asymptotically, similar to the part following (8),

$$\log D(X) \leq (n-1)\log\left(\frac{X}{n-1} + 1\right) + O(n).$$

Since π satisfies (5), we have $np + (n-1)\log\left(\frac{X}{n-1} + 1\right) + O(n) \geq n\log n - \Theta(n)$. Hence, $X \geq n^2/(2^p)\Theta(1) = \Omega(n^2/2^p)$, where the last equality holds since $p \leq \log n - \log\log n$ and hence $n^2/2^p \geq n\log n$. ■

By Lemma 1 at least a $(1 - 1/n)$-fraction of all permutations π require that high complexity. This shows that the running time of the Dobosiewicz sort is as stated in the theorem. The lower bound on Shaker sort has a very similar proof, with the proviso that we require $2n$ bits to encode one pass of the algorithm rather than n bits. This results in the claimed lower bound of $\Omega(n^2/4^p)$ (which is nan-vacuous only for for $p \leq \frac{1}{2}(\log n - \log\log n)$). ■

6. Sorting with queues and stacks

Knuth [11] and Tarjan [27] have studied the problem of sorting using a network of queues or stacks. The main variant of the problem is as follows: Given that the stacks or queues are arranged sequentially as shown in Figure 1, or in parallel as shown in Figure 2. Question: how many stacks or queues are needed to sort n elements with comparisons only? We assume that the input sequence is scanned from left to right, and the elements follow the arrows to go to the next stack or queue or output. In [12, 14] only the average-case analyses of the above two main variants was given, although the technique applies more in general to arbitrary acyclic networks of stacks and queues as studied in [27].

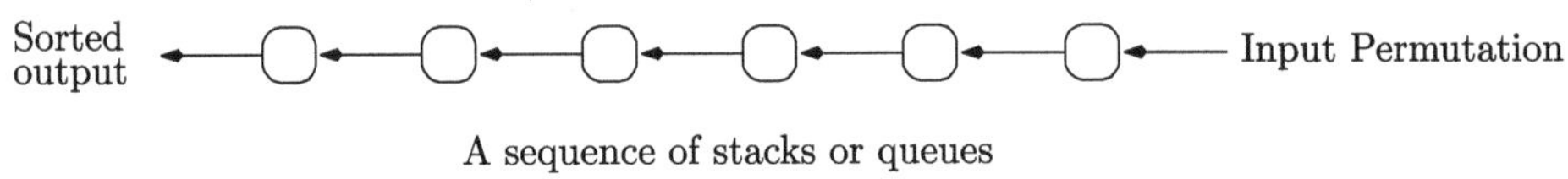

Fig. 1. Six stacks/queues arranged in sequential order

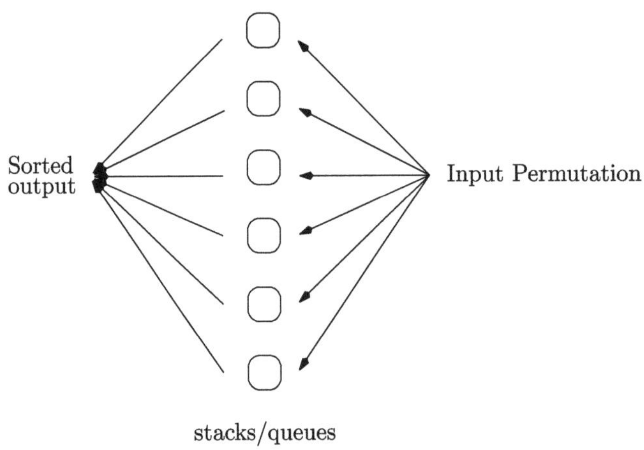

Fig. 2. Six stacks/queues arranged in parallel order

6.1. Sorting with sequential stacks

The sequential stack sorting problem is given in [11] exercise 5.2.4–20. We have k stacks numbered $S_0, \ldots, S_{k-1}$. The input is a permutation π of the elements $1, \ldots, n$. Initially we push the elements of π on S_0, at most one at a time, in the order in which they appear in π. At every step we can pop a stack (the popped elements will move left in Figure 1) or push an incoming element on a stack. The question is how many stack are needed for sorting π. It is known that $k = \log n$ stacks suffice, and $\frac{1}{2}\log n$ stacks are necessary in the worst-case [11, 27]. Here we prove that the same lower bound also holds on the average, using a very simple incompressibility argument.

Theorem 4. *On average (uniform distribution), at least* $\frac{1}{2}\log n$ *stacks are needed for sequential stack sort.*

Proof. Fix a random permutation π such that

$$C(\pi \mid n, P) \geq \log n! - \log n = n \log n - O(n),$$

where P is an encoding program to be specified in the following.

Assume that k stacks are sufficient to sort π. We now encode such a sorting process. For every stack, exactly n elements pass through it. Hence we need perform precisely n pushes and n pops on every stack. Encode a push as 0 and a pop as 1. It is easy to prove that different permutations must have different push/pop sequences on at least one stack. Thus with $2kn$ bits, we can completely specify the input permutation π. Then, as before,

$$2kn \geq \log n! - \log n = n \log n - O(n).$$

Therefore, we have $k \geq \frac{1}{2}\log n - O(1)$ for the random permutation π.

Since most (a $(1-1/n)$th fraction) permutations are incompressible, we calculate the average-case lower bound as:

$$\frac{1}{2}\log n \cdot \frac{n-1}{n} + 1 \cdot \frac{1}{n} \approx \frac{1}{2}\log n. \qquad \blacksquare$$

6.2. Sorting with parallel stacks

Clearly, the input sequence $2, 3, 4, \ldots, n, 1$ requires $n-1$ parallel stacks to sort. Hence the worst-case complexity of sorting with parallel stacks, as shown in Figure 2, is $n-1$. However, most sequences do not need this many stacks to sort in a parallel arrangement. The next two theorems show that on average, $\Theta(\sqrt{n})$ stacks are both necessary and sufficient. Observe that the result is actually implied by the connection between sorting with parallel stacks and *longest increasing subsequences* in [27], and the bounds on the length of longest increasing subsequences of random permutations given in [9, 17, 8]. However, the proofs in [9, 17, 8] use deep results from probability theory (such as Kingman's ergodic theorem) and are quite sophisticated. Here we give simple proofs using incompressibility arguments.

Theorem 5. *On average (uniform distribution), the number of parallel stacks needed to sort n elements is $O(\sqrt{n})$.*

Proof. Consider an incompressible permutation π satisfying

$$C(\pi \mid n) \geq \log n! - \log n. \tag{9}$$

We use the following trivial algorithm (described in [27]), to sort π with stacks in the parallel arrangement shown in Figure 2. Assume that the stacks are $S_0, S_1, \ldots$, and the input sequence is denoted as $x_1, \ldots, x_n$.

Algorithm Parallel-Stack-Sort

1. For $i = 1$ to n do

   Scan the stacks from left to right, and push x_i on the the first stack S_j whose top element is larger than x_i. If such a stack doesn't exist, put x_i on the first empty stack.

2. Pop the stacks in the ascending order of their top elements.

We claim that algorithm Parallel-Stack-Sort uses $O(\sqrt{n})$ stacks on the permutation π. First, we observe that if the algorithm uses m stacks on π then we can identify an increasing subsequence of π of length m as in [27]. This can be done by a trivial backtracking starting from the top element of the last stack. Then we argue that π cannot have an increasing subsequence of length longer than $e\sqrt{n}$, where e is the natural constant, since it is compressible by at most $\log n$ bits.

Suppose that σ is a longest increasing subsequence of π and $m = |\sigma|$ is the length of σ. Then we can encode π by specifying:

1. a description of this encoding scheme in $O(1)$ bits;
2. the number m in $\log m$ bits;
3. the combination σ in $\log \binom{n}{m}$ bits;
4. the locations of the elements of σ in π in at most $\log \binom{n}{m}$ bits; and
5. the remaining π with the elements of σ deleted in $\log(n-m)!$ bits.

This takes a total of

$$\log(n-m)! + 2\log\frac{n!}{m!(n-m)!} + \log m + O(1) + 2\log\log m$$

bits, where the last $\log\log m$ term serves to self-delimitingly encode m. Using Stirling's approximation, and the fact that $\sqrt{n} \leq m = o(n)$, the above expression is upper bounded by:

$$\log n! + \log\frac{(n/e)^n}{(m/e)^{2m}\big((n-m)/e\big)^{n-m}} + O(\log n)$$

$$\approx \log n! + m\log\frac{n}{m^2} + (n-m)\log\frac{n}{n-m} + m\log e + O(\log n)$$

$$\approx \log n! + m\log\frac{n}{m^2} + 2m\log e + O(\log n)$$

This description length must exceed the complexity of the permutation which is lower-bounded in (9). Therefore, approximately $m \leq e\sqrt{n}$, and hence $m = O\big(\sqrt{n}\big)$. Hence, the average complexity of Parallel-Stack-Sort is

$$O\big(\sqrt{n}\big)\cdot\frac{n-1}{n} + n\cdot\frac{1}{n} = O\big(\sqrt{n}\big). \qquad \blacksquare$$

Theorem 6. *On average (uniform distribution), the number of parallel stacks required to sort a permutation is* $\Omega\big(\sqrt{n}\big)$.

Proof. Let A be a sorting algorithm using parallel stacks. Fix a random permutation π with $C(\pi \mid n, P) \geq \log n! - \log n$, where P is the program to do the encoding discussed in the following. Suppose that A uses T parallel stacks to sort π. This sorting process involves a sequence of moves, and we can encode this sequence of moves by a sequence of instructions of the types:

- push to stack i,
- pop stack j,

where the element to be pushed is the next unprocessed element from the input sequence, and the popped element is written as the next output element. Each of these term requires $\log T$ bits. In total, we use precisely $2n$ terms since every element has to be pushed once and has to be popped once. Such a sequence is unique for every permutation.

Thus we have a description of an input sequence in $2n \log T$ bits, which must exceed $C(\pi \mid n, P) \geq n \log n - O(\log n)$. It follows that $T \geq \sqrt{n} = \Omega(\sqrt{n})$.

This yields the average-case complexity of A:

$$\Omega(\sqrt{n}) \cdot \frac{n-1}{n} + 1 \cdot \frac{1}{n} = \Omega(\sqrt{n}). \qquad \blacksquare$$

6.3. Sorting with parallel queues

It is easy to see that sorting cannot be done with a sequence of queues. So we consider the complexity of sorting with parallel queues. It turns out that all the result in the previous subsection also hold for queues.

As noticed in [27], the worst-case complexity of sorting with parallel queues is n, since the input sequence $n, n-1, \ldots, 1$ requires n queues to sort. We show in the next two theorems that on average, $\Theta(\sqrt{n})$ queues are both necessary and sufficient. Again, the result is implied by the connection between sorting with parallel queues and longest *decreasing* subsequences given in [27] and the bounds in [9, 17, 8] (with sophisticated proofs). Our proofs are trivial given the proofs in the previous subsection.

Theorem 7. *On average (uniform distribution), the number of parallel queues needed to sort n elements is upper bounded by $O(\sqrt{n})$.*

Proof. The proof is very similar to the proof of Theorem 5. We use a slightly modified greedy algorithm as described in [27]:

Algorithm Parallel-Queue-Sort

1. For $i = 1$ to n do

   Scan the queues from left to right, and append x_i on the the first queue whose rear element is smaller than x_i. If such a queue doesn't exist, put x_i on the first empty queue.

2. Delete the front elements of the queues in the ascending order.

Again, we claim that algorithm Parallel-Queue-Sort uses $O(\sqrt{n})$ queues on every permutation π, that cannot be compressed by more than $\log n$ bits. We first observe that if the algorithm uses m queues on π then a decreasing subsequence of π of length m can be identified, and we then argue that π cannot have a decreasing subsequence of length longer than $e\sqrt{n}$, in a way analogous to the argument in the proof of Theorem 5. $\blacksquare$

Theorem 8. *On average (uniform distribution), the number of parallel queues required to sort a permutation is* $\Omega(\sqrt{n})$.

Proof. The proof is the same as the one for Theorem 6 except that we should replace "push" with "enqueue" and "pop" with "dequeue". ■

REFERENCES

[1] B. Brejóva, Analyzing variants of Shellsort, *Information Processing Letters,* **79:5** (2001), 223–227.

[2] H. Buhrman, T. Jiang, M. Li and P. Vitanyi, New applications of the incompressibility method: Part II, *Theoretical Computer Science,* **235:1** (2000), 59–70.

[3] W. Dobosiewicz, An efficient variant of bubble sort, *Information Processing Letters,* **11:1** (1980), 5–6.

[4] R. W. Floyd, Algorithm 245: Treesort 3. *Communications of the ACM,* **7** (1964), 701.

[5] J. Incerpi and R. Sedgewick, Improved upper bounds on Shellsort, *Journal of Computer and System Sciences,* **31** (1985), 210–224.

[6] J. Incerpi and R. Sedgewick, Practical variations of Shellsort, *Information Processing Letters,* **26:1** (1980), 37–43.

[7] S. Janson and D. E. Knuth, Shellsort with three increments, *Random Struct. Alg.,* **10** (1997), 125–142.

[8] S. V. Kerov and A. M. Versik, Asymptotics of the Plancherel measure on symmetric group and the limiting form of the Young tableaux, *Soviet Math. Dokl.*, **18** (1977), 527–531.

[9] J. F. C. Kingman, The ergodic theory of subadditive stochastic processes, *Ann. Probab.,* **1** (1973), 883–909.

[10] A. N. Kolmogorov, Three approaches to the quantitative definition of information, *Problems Inform. Transmission,* **1:1** (1965), 1–7.

[11] D. E. Knuth, *The Art of Computer Programming, Vol. 3: Sorting and Searching,* Addison-Wesley, 1973 (1st Edition), 1998 (2nd Edition).

[12] T. Jiang, M. Li and P. Vitanyi, Average complexity of Shellsort (preliminary version), *Proc. ICALP99,* Lecture Notes in Computer Science, Vol. 1644, Springer-Verlag (Berlin, 1999), pp. 453–462.

[13] T. Jiang, M. Li and P. Vitanyi, A lower bound on the average-case complexity of Shellsort, *J. Assoc. Comp. Mach.,* **47:5** (2000), 905–911.

[14] T. Jiang, M. Li and P. Vitanyi, Average-case analysis of algorithms using Kolmogorov complexity, *Journal of Computer Science and Technology,* **15:5** (2000), 402–408.

[15] T. Jiang, M. Li and P. Vitányi, The average-case area of Heilbronn-type triangles, *Random Structures and Algorithms,* **20:2** (2002), 206–219.

[16] M. Li and P. M. B. Vitányi, *An Introduction to Kolmogorov Complexity and its Applications,* Springer-Verlag, 2nd Edition (New York, 1997).

[17] B. F. Logan and L. A. Shepp, A variational problem for random Young tableaux, *Advances in Math.,* **26** (1977), 206–222.

[18] I. Munro, Personal communication, 1992.

[19] A. Papernov and G. Stasevich, A method for information sorting in computer memories, *Problems Inform. Transmission,* **1:3** (1965), 63–75.

[20] B. Poonen, The worst-case of Shellsort and related algorithms, *J. Algorithms,* **15:1** (1993), 101–124.

[21] C. G. Plaxton, B. Poonen and T. Suel, Improved lower bounds for Shellsort, in: *Proc. 33rd IEEE Symp. Foundat. Comput. Sci.* (1992), pp. 226–235.

[22] V. R. Pratt, *Shellsort and Sorting Networks,* Ph.D. Thesis, Stanford Univ. (1972).

[23] R. Schaffer and R. Sedgewick, *J. Algorithms,* **15** (1993), 76–100.

[24] R. Sedgewick, Analysis of Shellsort and related algorithms, presented at the *Fourth Annual European Symposium on Algorithms* (Barcelona, September, 1996).

[25] R. Sedgewick, Open problems in the analysis of sorting and searching algorithms, Presented at *Workshop on Prob. Analysis of Algorithms* (Princeton, 1997).

[26] D. L. Shell, A high-speed sorting procedure, *Commun. ACM,* **2:7** (1959), 30–32.

[27] R. E. Tarjan, Sorting using networks of queues and stacks, *Journal of the ACM,* **19** (1972), 341–346.

[28] M. A. Weiss and R. Sedgewick, Bad cases for Shaker-sort, *Information Processing Letters,* **28:3** (1988), 133–136.

[29] J. W. J. Williams *Comm. ACM,* **7** (1964), 347–348.

[30] A. C. C. Yao, An analysis of $(h, k, 1)$-Shellsort, *J. of Algorithms,* **1** (1980), 14–50.

Paul Vitányi
CWI, Kruislaan 413
1098 SJ Amsterdam
The Netherlands

paulv@cwi.nl

BOLYAI SOCIETY
MATHEMATICAL STUDIES, 16

Entropy, Search, Complexity, pp. 233–264.

Recognition Problems in Combinatorial Search

GÁBOR WIENER*

We survey some recent results concerning recognition problems. Recognition problems are special combinatorial search problems, where we need not find the hidden element x itself, just compute the value $f(x)$ for a given function f. We mainly use the approaches of classical search theory and two-party deterministic communication complexity.

1. Introduction

The classical problem of search theory is to find a hidden element $x \in H$ by asking as few questions of type "is $x \in A$?" as possible, where S is a finite underlying set and $\mathcal{R}$ is an arbitrary set system on S. The *recognition problem* is somewhat similar: now a function $f : H \to \{0,1\}$ is also given and we need not find x itself, just have to compute the value $f(x)$.

Though this model of search arises quite naturally, it is rarely treated in full generality. However, some special cases are examined thoroughly. The most studied special case is the recognition of graph properties. This case is not examined here, the interested reader is referred to the books by Bollobás [2], Yap [15], or Aigner [1]. Another special case (as we will see) is (two-party, deterministic) communication complexity. The relationship between communication and recognition complexity is treated in [13].

We use the approaches and methods of both "classical" search theory and communication complexity to solve recognition problems, so it is worth reviewing the most basic notions and theorems of these topics in a nutshell. This is the subject of the next section.

*Research supported in part by OTKA Grants T029255 and T034702.

The third section contains the basic notions, definitions, and some general theorems, while the fourth one analyzes the connection between communication and recognition complexity by reviewing the so-called generalized communication model. Further analysis of the communication model can be found in [13]. In the fifth section some interesting special cases concerning partially ordered sets are treated. Up to this point we are dealing with *adaptive problems,* i.e. problems, where the next question may depend on the answers to the previous ones. In the sixth section the predetermined recognition problem is studied, where the questions are all fixed beforehand. The last section is dedicated to the open problems.

2. Preliminaries

2.1. Combinatorial search

For a detailed survey of combinatorial search see the books by Aigner [1] and Du and Hwang [3] and the papers of Katona [5], [6] and Rényi [12].

Search problems. Let S be a finite, non-empty set, called the *search domain,* $x \in S$, and let $\mathcal{F}$ be a family of functions on S, called the *test family.* We choose a function $f_1 \in \mathcal{F}$ and receive as answer the value $f_1(x)$. With this information we choose again a function $f_2 \in \mathcal{F}$ and get back the value $f_2(x)$, and so on. A successful search algorithm $\mathcal{A}$ consists in the choice of functions $f_1, f_2, f_3, \ldots \in \mathcal{F}$ such that the values $f_1(x), f_2(x), f_3(x), \ldots$ determine x uniquely. We tacitly assume that at least one such sequence always exists. The pair $(S, \mathcal{F})$ is called a *search problem.*

Since S is finite, $f(S) \subseteq \{0, 1, \ldots, q-1\}$, for every function $f \in \mathcal{F}$, for some $q \geq 2$. Now $(S, \mathcal{F})$ is called an (n, q)*-problem,* where $n = |S|$. If $q = 2$, then we speak of a *binary search problem.* We focus our attention to binary problems.

A search algorithm is called *adaptive* or *sequential* (or sometimes *dynamic*) if the choice of a function may depend on the values obtained until then. If all functions are fixed beforehand, then the algorithm is called *non-adaptive* or *predetermined* (or sometimes *static*). Since predetermined algorithms are special cases of sequential ones, they take no shorter than the best sequential one.

Our aim is to minimize the length of successful algorithms in some sense. If we simply consider all possibilities for x, then we speak of a *worst-case problem*. We may be interested in minimizing the average length given a certain distribution, this is called an *average-case problem*. The tests may have different costs and we may look for algorithms of minimal cost or a combination of time and cost (so-called *trade-off problems*). In this paper we always consider worst-case problems.

Let $(S, \mathcal{F})$ be a search problem and let $\mathcal{A} = \big(f_1, f_2, \ldots, f_{l(x)}\big)$ be a successful algorithm. Suppose that

$$\big\{y \mid y \in S,\ f_1(y) = f_1(x), f_2(y) = f_2(x), \ldots, f_{l(x)-1}(y) = f_{l(x)-1}(x)\big\} \supsetneq \{x\}.$$

Then

- the number $l(x)$ is called the (search) *length* for x in $\mathcal{A}$,
- $L(\mathcal{A}) = \max\limits_{x \in S} l(x)$ is called the *length* of algorithm $\mathcal{A}$,
- the number $\min\limits_{\mathcal{A}} L(\mathcal{A})$ is called the (worst-case) (adaptive) *cost* or (worst-case) *complexity* of the adaptive search problem $(S, \mathcal{F})$ and is denoted by $L(S, \mathcal{F})$,
- the number $\min\limits_{\mathcal{A}\text{ non-ad.}} L(\mathcal{A})$ is called the (worst-case) non-adaptive complexity of the non-adaptive search problem $(S, \mathcal{F})$ and is denoted by $L_{\text{pre}}(S, \mathcal{F})$.

An adaptive algorithm $\mathcal{A}$ is said to be *optimal* if $L(\mathcal{A}) = L(S, \mathcal{F})$.
A non-adaptive algorithm $\mathcal{A}$ is said to be *optimal* if $L(\mathcal{A}) = L_{\text{pre}}(S, \mathcal{F})$.

Our knowledge after tests $f_1, f_2, \ldots, f_k$ and answers $e_1, e_2, \ldots, e_k$, respectively is that the hidden element x belongs to the set

$$\big\{y \mid y \in S,\ f_1(y) = e_1,\ f_2(y) = e_2,\ \ldots,\ f_k(y) = e_k\big\}.$$

This set is denoted by $S(e_1, e_2, \ldots, e_k)$. Let furthermore

$$S_i(x) = S\big(f_1(x), f_2(x), \ldots, f_r(x)\big).$$

Now it is clear that

$$S = S_0(x) \supseteq S_1(x) \supseteq S_2(x) \supseteq S_2(x) \supseteq \ldots \supseteq S_{l(x)}(x) = \{x\}.$$

Search trees. The most useful ways to represent search algorithms are *search trees* (also called *decision trees*).

An (n,q)-tree is a rooted tree with n leaves, where every inner node has at most q sons. The length of a leaf l is the length of the unique path from the root to l. The depth of a rooted tree T is denoted by $L(T)$.

If $q = 2$, then we speak of a binary tree.

Let $\mathcal{A} = (f_1, f_2, \ldots)$ be an algorithm for the (n,q)-problem $(S, \mathcal{F})$. By associating the root of a rooted tree to S and nodes of the rooted tree to all non-empty sets $S(e_1, e_2, \ldots, e_k)$, we obtain an (n,q)-tree T, whose leaves x' correspond bijectively to the elements $x \in S$ and the length of the leaf x' in the tree is the same as the length of the element x in $\mathcal{A}$, thus $L(\mathcal{A}) = L(T)$. The rooted tree T is called the *decision tree* corresponding to $\mathcal{A}$. It is clear that the set $S(e_1, e_2, \ldots, e_k)$ associated to an inner node v is the set of the elements x associated to the leaves x' that can be reached from node v. The number of leaves in a decision tree corresponding to any algorithm $\mathcal{A}$ of an (n,q)-problem is n, thus the depth of the tree is at least $\log_q n$. Since $L(T) = L(\mathcal{A})$, we have the following inequality, which is called the *information theoretic lower bound:*

Proposition 2.1. *If $(S, \mathcal{F})$ is an (n,q)-problem, then*

$$L(S, \mathcal{F}) \geq \lceil \log_q n \rceil.$$

Binary search. The binary case is the most important and most frequently studied special case of search problems. The corresponding decision trees are binary trees. For an $(n, 2)$-problem the question "what is $f(x)$?" is equivalent to the question "does x belong to $R = \{y \mid f(y) = 1\}$?". If we receive yes, as an answer, then $f(x) = 1$, otherwise $f(x) = 0$. Further on, for binary problems an algorithm will be a sequence of sets, and binary problems will be denoted by $(S, \mathcal{R})$. The set system $\mathcal{R}$ is called the *test family,* and the elements of $\mathcal{R}$ are called *question sets.*

A successful algorithm $\mathcal{A} = (R_1, R_2, \ldots, R_k)$ together with the sequence of answers (yes or no) determines the unknown element x. That is,

$$\forall y \in S \; \exists i \leq k : x \in R_i,\; y \notin R_i \text{ or } x \notin R_i,\; y \in R_i.$$

If this condition holds, we say that x is *separated* from the other elements of S. In the predetermined case clearly every element must be separated from the others.

$\mathcal{H} \subseteq 2^S$ is called a *separating set system* on S if for any $x, y \in S$, $x \neq y$ $\exists R \in \mathcal{H} : x \in R,\; y \notin R$ or $x \notin R,\; y \in R$.

Hence binary predetermined algorithms and separating set systems are the same. Separating systems are also important in adaptive problems: if $(S, \mathcal{R})$ is a binary adaptive problem, then $\mathcal{R}$ must be a separating system on S.

2.2. Communication complexity

For a detailed treatment the reader is referred to the book by Nisan and Kushilevitz [7] and the surveys of Lovász [9] and Lengauer [8].

Yao's two-party model. Let X, Y, Z be arbitrary finite sets and let $f : X \times Y \to Z$ be an arbitrary function. There are two players, Alice and Bob, who wish to compute the value $f(x, y)$, for some inputs $x \in X$, $y \in Y$. The difficulty is that Alice only knows x and Bob only knows y, thus they need to communicate with each other. The communication will be carried out according to a so-called *protocol* (something like an algorithm), which depends only on f. This protocol consists of the players sending bits to each other until $f(x, y)$ is determined.

At each stage, the protocol P must determine whether the run terminates; if the run has terminated, the protocol must determine $f(x, y)$, and if the run has not terminated P must specify which player sends the next bit. These informations may depend only on P and the bits communicated so far, since this is the only common knowledge of Alice and Bob. The protocol also determines what the player in turn should send; this depends on not only the bits exchanged so far but also on his/her input x or y. The *cost* of a protocol P is the maximal number of bits communicated on the possible inputs $(x, y) \in X \times Y$. The *complexity* of a function f is the minimum cost of a protocol that computes f (i.e. for every $(x, y) \in X \times Y$ computes $f(x, y)$). We denote the (deterministic) communication complexity of f by $D(f)$.

Matrices and rectangles. With every function $f : X \times Y \to Z$ we may associate a matrix M_f of dimensions $|X| \times |Y|$. This matrix not only helps us to understand the nature of protocols better but plays a most important role in some bounds.

Let $f : X \times Y \to Z$ be a function. Let the rows and the columns of the matrix M_f of dimensions $|X| \times |Y|$ be indexed with the elements of X

and Y, respectively and let the (x, y) entry of M_f be $f(x, y)$. This matrix is called the *communication matrix* of f.

Let M be the communication matrix of the function $f : X \times Y \to Z$. Let us see what happens in the matrix during the run of a protocol. Assume, without loss of generality that the first bit is sent by Alice. This bit is determined by only the protocol and her input (some row index i), thus the set of rows of M can be divided into two parts: a part that contains rows for which 0 is sent and a part that contains the other rows, for which 1 is sent. In this way M is divided into two submatrices M_0 and M_1. Notice that the matrices M_0 and M_1 are determined by the protocol and the bit Alice sends determines only in which part her row lies. The next bit divides submatrix M_0 or M_1 into two submatrices, and so on. Since the players' common knowledge after some bits sent is that (x, y) lies in the current submatrix, the protocol terminates if and only if the elements of this current submatrix are all the same. Such a matrix is called *monochromatic.*

A *combinatorial rectangle* (in short, a *rectangle*) in $X \times Y$ is a subset $R \subseteq X \times Y$ such that $R = A \times B$ for some $A \subseteq X$ and $B \subseteq Y$. Rectangles can also be considered as submatrices of the communication matrix.

The following theorem gives a strategy for proving lower bounds on the communication complexity of a function.

Theorem 2.2. *If every partition of $X \times Y$ into f-monochromatic rectangles contains at least t rectangles, then $D(f) \geq \lceil \log t \rceil$.*

Computing the size of the least partition of $X \times Y$ into monochromatic rectangles is often complicated. The notion of fooling sets serves to handle this problem.

Let $f : X \times Y \to \{0, 1\}$. A set $S \subseteq X \times Y$ is called a *fooling set* for f, if for every two distinct pairs $(x_1, y_1) \in S$ and $(x_2, y_2) \in S$ either

- $f(x_2, y_2) \neq f(x_1, y_1)$ or
- $f(x_1, y_2) \neq f(x_1, y_1)$ or
- $f(x_2, y_1) \neq f(x_1, y_1)$.

Theorem 2.3 (Fooling set method). *If a fooling set S of size t exists for function f, then $D(f) \geq \lceil \log t \rceil$.*

Algebraic bounds. We have only seen combinatorial type bounds up to this point. Now we mention some algebraic techniques, based on the rank of the communication matrix.

Theorem 2.4 (Mehlhorn and Schmidt [11]). *For any function* $f : X \times Y \to \{0,1\}$,

$$D(f) \geq \log rk(M_f) + 1.$$

The next bound is quite obvious and not too strong, nevertheless it is the best one amongst the rank type upper bounds.

Proposition 2.5. $D(f) \leq rk(M_f) + 1$.

Simultaneous protocols. In the definition of communication complexity Alice and Bob alternate sending messages to each other. We may ask how the complexity changes if the interaction between the players (that is, the number of alternations) is limited.

A k-round protocol is a protocol, where on every input there are at most k alternations between bits sent by Alice and bits sent by Bob. For example, a one-round protocol is a protocol, where Alice sends a message (containing an arbitrary number of bits) to Bob, and Bob sends back the answer. The k-round communication complexity of a function f, denoted by $D_k^A(f)$, is the cost of the best k-round protocol that computes f, where Alice sends the first message. We use $D_k^B(f)$ to denote the cost of the best k-round protocol where Bob sends the first message.

The one-round complexities are easy to compute:

Proposition 2.6. *Let* $f : X \times Y \to \{0,1\}$ *be an arbitrary Boolean function. Let furthermore* a *and* b *be the number of distinct rows and columns in the communication matrix* M_f*, respectively. Then* $D_1^A(f) = \lceil \log a \rceil + 1$, $D_1^B(f) = \lceil \log b \rceil + 1$.

Finally, we define an even more restricted type of protocols.

A protocol, where the players send messages not to each other, but to an independent referee, and this referee computes the result, is called a *simultaneous protocol.* The *simultaneous communication complexity* of a function f, denoted by $D_1(f)$, is the cost of the best simultaneous protocol that computes f.

This model eliminates interaction completely. The simultaneous complexity of a function can be easily determined:

Proposition 2.7. $D_1(f) = D_1^A(f) + D_1^B(f)$.

3. Recognition problems

Now we are ready to start studying recognition problems. First we introduce some useful notions, then we give the basic definitions used in recognition complexity.

3.1. Notions and definitions

Let S be a finite underlying set and $\mathcal{H} \subseteq 2^S$ be an arbitrary set system. A set $H \in \mathcal{H}$ is called a *minimal set* of $\mathcal{H}$ if there is no non-empty set $G \in \mathcal{H}$ such that $G \subsetneq H$. H is *maximal* if there is no set $G \in \mathcal{H}$, such that $H \subsetneq G$. The set system of minimal and maximal sets of $\mathcal{H}$ is denoted by $\min \mathcal{H}$ and $\max \mathcal{H}$, respectively.

Let $\mathcal{H}$ be a set system on the underlying set S. We define set systems

$$\mathcal{H}^{k\cap} = \big\{ \cap H_i \mid H_i \in \mathcal{A} \subseteq \mathcal{H},\ |\mathcal{A}| = k \big\}, \quad \text{for} \quad k \in \mathbb{N},$$

$$\mathcal{H}^{\cap} = \bigcup_{i \in \mathbb{N}} \mathcal{H}^{i\cap},$$

$$\mathcal{H}^{k\cup} = \big\{ \cup H_i \mid H_i \in \mathcal{A} \subseteq \mathcal{H},\ |\mathcal{A}| = k \big\}, \quad \text{for} \quad k \in \mathbb{N},$$

$$\mathcal{H}^{\cup} = \bigcup_{i \in \mathbb{N}} \mathcal{H}^{i\cup},$$

$$\mathcal{H}^{-} = \{ \overline{H} \mid H \in \mathcal{H} \},$$

and

$$\mathcal{H}|_A = \{ H \cap A \mid H \in \mathcal{H} \}, \quad \text{for} \quad A \subseteq S.$$

Let $f : A \to B$ be a function. The set $\big\{ f(s) \mid s \in S \subseteq A \big\}$ is denoted by $f(S)$ and the set $\big\{ a \mid f(a) = b \big\}$ is denoted by $f^{-1}(b)$.

We have seen in the Introduction informally what a recognition problem is. Now we give the precise definition.

Definition 3.1. The triple $(S, \mathcal{R}, f)$ is called a *recognition problem,* if S is a finite set, $\mathcal{R}$ is a set system on S, and f is a function from S to $\{0, 1\}$. The elements of $\mathcal{R}$ are called question sets.

The definition of *recognition algorithms* is similar to the definition of binary search algorithms, the only difference being that we do not have to determine the hidden element itself, just have to recognize whether the value $f(x)$ is 0 or 1.

Definition 3.2. Let $(S, \mathcal{R}, f)$ be a recognition problem, $x \in S$ is a hidden element. We choose a set $R_1 \in \mathcal{R}$ and receive as answer $\left|R_1 \cap \{x\}\right|$, that is, 1, if $x \in R_1$ and 0, if $x \notin R_1$. With this information we choose again a set $R_2 \in \mathcal{R}$ and get back the number $\left|R_2 \cap \{x\}\right|$, and so on. A successful recognition algorithm $\mathcal{A}$ consists in the choice of sets $R_1, R_2, R_3, \ldots \in \mathcal{R}$ such that the answers determine $f(x)$ uniquely. We tacitly assume that at least one such sequence always exists.

Just like for search algorithms, a recognition algorithm is called *adaptive* or *sequential* if the choice of a set may depend on the answers obtained until then (as in the previous definition). If all sets are fixed beforehand, then the recognition algorithm is called *non-adaptive* or *predetermined.* Predetermined recognition algorithms are special cases of sequential ones, so they take no shorter than the best sequential one, just as for search algorithms. We are dealing with adaptive algorithms, unless the opposite is declared.

Definition 3.3. Let $(S, \mathcal{R}, f)$ be a recognition problem and let $\mathcal{A} = \big(R_1, R_2, \ldots, R_{r(x)}\big)$ be a successful recognition algorithm. Suppose that the answers to the first $r(x) - 1$ questions sets do not determine $f(x)$ uniquely. Then

- the number $r(x)$ is called the *recognition length* for x in $\mathcal{A}$,
- $R(\mathcal{A}) = \max\limits_{x \subset S} r(x)$ is called the *recognition length* of the algorithm $\mathcal{A}$,
- the number $\min\limits_{\mathcal{A}} R(\mathcal{A})$ is called the (adaptive) *recognition complexity* of the adaptive recognition problem $(S, \mathcal{R}, f)$ and is denoted by $g(S, \mathcal{R}, f)$,
- the number $\min\limits_{\mathcal{A} \text{ non-ad.}} R(\mathcal{A})$ is called the *non-adaptive recognition complexity* of the non-adaptive recognition problem $(S, \mathcal{R}, f)$ and is denoted by $g_{\text{pre}}(S, \mathcal{R}, f)$.

An adaptive recognition algorithm $\mathcal{A}$ is said to be *optimal* if $R(\mathcal{A}) = g(S, \mathcal{R}, f)$.
A non-adaptive algorithm $\mathcal{A}$ is said to be *optimal* if $R(\mathcal{A}) = g_{\text{pre}}(S, \mathcal{R}, f)$.

If S and $\mathcal{R}$ are fixed, then $g(S, \mathcal{R}, f)$ is simply denoted by $g(f)$.

Since the question "is $x \in A$?" is equivalent to the question "is $x \in \overline{A}$?", for convenience's sake it is assumed, that the following condition holds:

(C) $\mathcal{R}$ is complementation-closed.

Now let us see what do we know after the question sets $B_1, B_2, \ldots, B_k$ were asked. Every answer is of the form $x \in B_i'$ for $i = 1, 2, \ldots, k$, where B_i' is either B_i or $\overline{B_i}$, therefore by condition (C), $B_i' \in \mathcal{R}$. Thus our knowledge is that the unknown element x belongs to the set $T = \cap_{i=1}^{k} B_i'$, and clearly $T \in \mathcal{R}^{k\cap}$. Therefore the sequence $(B_1, B_2, \ldots, B_k)$ is a succesful recognition algorithm if and only if the function f is constant on the set T (otherwise we could not decide whether $f(x)$ is 0 or 1). Such a set is called *f-monochromatic* or simply *monochromatic.* That is, a set is monochromatic if and only if it is a subset of either $f^{-1}(0)$ or $f^{-1}(1)$. Since $T \in \mathcal{R}^{k\cap}$ for some k, T must be a member of $\mathcal{R}^{\cap}$.

The assumption that at least one succesful recognition algorithm exists means that the elements of $\min \mathcal{R}^{\cap}$ must be monochromatic. In fact, we cannot distinguish between two elements of such a minimal set, so it is worth assuming that

(M) $$H \in \min \mathcal{R}^{\cap} \Rightarrow |H| = 1.$$

Conditions (C) and (M) imply that $\mathcal{R}$ is a separating system (in fact, $\mathcal{R}$ is a *completely separating system* then, i.e. $\forall x, y \in S : \exists R \in \mathcal{R} : x \in R, y \notin R$).

The search tree (or decision tree) corresponding to a recognition algorithm is defined similarly as for classic search algorithms (see Search trees, page 235). The only difference is that the leaves here are associated to monochromatic sets of elements.

Definition 3.4. Let $\mathcal{B}$ be an arbitrary set system on S and $H \subseteq S$ be an arbitrary set. Let furthermore $\mathcal{B}_H$ be the set system of those elements of $\mathcal{B}$ that contain set H. The *span* of H in $\mathcal{B}$, denoted by $sp_{\mathcal{B}}(H)$ is the set system $\min \mathcal{B}_H$.

Definition 3.5. The *coordinate set* of an element x in set system $\mathcal{R}$ is $sp_{\mathcal{R}}(\{x\})$. It is denoted by $\mathcal{C}_{\mathcal{R}}(x)$ or simply by $\mathcal{C}(x)$. The elements of the coordinate set are called *coordinates.*

Remark. By this definition, coordinates are subsets of S and the coordinate set is a set system on S. Note that a coordinate set is never empty (unless $\mathcal{R}$ itself is empty), by condition (C).

Now we show that an element is determined by its coordinates, i.e. the name "coordinate" is legitimate.

Proposition 3.6.

$$\bigcap_{C \in \mathcal{C}(x)} C = \{x\}.$$

Proof. Consider the intersection of those sets of $\mathcal{R}$ that contain x. This intersection T is an element of $\min \mathcal{R}^{\cap}$, otherwise there would exist a non-empty set $U \in \mathcal{R}^{\cap}$, such that $U \subsetneq T$. Let $U = R_1 \cap R_2 \cap \ldots \cap R_j$, where $R_i \in \mathcal{R}$. If $x \in U$, then $x \in R_i$, for $i = 1, 2, \ldots, j$, thus $T \subseteq R_i$, for $i = 1, 2, \ldots, j$ by the definition of T. This implies $T \subseteq U$, a contradiction. On the other hand, if $x \notin U$, then there exists an index a such that $x \notin R_a$. Thus $x \in \overline{R_a}$, and $\overline{R_a} \in \mathcal{R}$, by condition (C). Therefore $T \subseteq \overline{R_a}$, i.e. the sets T and R_a are disjoint. Since U is a subset of R_a, the sets T and U are also disjoint, from which $U = \emptyset$ follows, again a contradiction.

Now, since T belongs to $\min \mathcal{R}^{\cap}$, it contains only x itself, by condition (M). The intersection of the coordinates of x contains x, and is obviously a subset of T, which proves the assertion. ■

Remark. In fact the name "coordinate" would be even better justified if the intersection of less than $|\mathcal{C}(x)|$ question sets could not be equal to $\{x\}$. However, this is not true in general, but the generalized communication model (see Section 4.2) shall satisfy this property.

3.2. General bounds

In this section we prove some general lower bounds on recognition complexity. A part of these bounds is obtained by generalizing certain communication complexity theorems. The reason why communication complexity methods can be used here is that communication complexity problems are special recognition problems, as we shall see it in Section 4.1. The methods of search theory are also used.

First we formulate the information theoretic lower bound for recognition problems.

Theorem 3.7. *Let $(S, \mathcal{R}, f)$ be a recognition problem. If the underlying set S cannot be partitioned into less then k f-monochromatic sets belonging to $\mathcal{R}^{\cap}$, then*

$$g(S, \mathcal{R}, f) \geq \lceil \log k \rceil.$$

Proof. The leaves of the search tree are f-monochromatic sets that belong to $\mathcal{R}^\cap$ and furthermore they partition S, thus their number cannot be less than k. Therefore the depth of the search tree is at least $\lceil \log k \rceil$. ■

Corollary 3.8. *Let μ be an arbitrary probability distribution on the underlying set S. If for every f-monochromatic set $R \in \mathcal{R}^\cap$ we have $\mu(R) \leq \delta$, then*

$$g(S, \mathcal{R}, f) \geq \left\lceil \log \frac{1}{\delta} \right\rceil .$$

Proof. Now the underlying set clearly cannot be partitioned into less then $1/\delta$ f-monochromatic sets of $\mathcal{R}^\cap$, thus the assertion follows from Theorem 3.7. ■

A special case of this proposition is the following

Corollary 3.9 (Fooling set method). *Let $H \subseteq S$ be a set with the property that for every two elements $x, y \in H$, there is no monochromatic set in $\mathcal{R}^\cap$ that contains both of them. (Such a set H is called a* fooling set *for f on $\mathcal{R}$.) Now*

$$g(S, \mathcal{R}, f) \geq \log |H|.$$

Proof. Let

$$\mu(x) = \begin{cases} 1/|H|, & \text{if } x \in H, \\ 0, & \text{if } x \notin H. \end{cases}$$

The μ-measure of any monochromatic set in $\mathcal{R}^\cap$ is at most $1/|H|$, by the condition on H. Now the assertion follows from Corollary 3.8. ■

Theorem 3.7. and Corollary 3.9. are generalizations of Theorem 2.2. and Theorem 2.3., respectively. This shall be proved in Section 4.1., where we show that deterministic two-party communication is a special case of recognition problems. The question arises naturally, whether there are other communication complexity theorems that can be generalized. In order to answer this question, we should analyze what is behind these bounds. Do they require some special feature that only communication problems possess or not? If yes, how special? The answers for all these questions will be given in Section 4. For the moment, we mention that most of the communication complexity bounds will not be true in the general setting of recognition problems, nevertheless we shall see models where a great number of them will be possible to generalize.

Now we examine another quite natural question: which functions and set systems produce extremal values of complexity. For a given non-constant function f it is easy to choose a set system $\mathcal{R}$, over which the complexity of the function is small: if $\mathcal{R}$ contains $f^{-1}(0)$, then $g(S,\mathcal{R},f) = 1$. Thus over the set system 2^S, every function has complexity 1 (except constant functions that have complexity 0 over any set system). To find a set system, over which the complexity of a given non-constant function f is large is not difficult, either. Let $y \in S$ be an element of the smaller (not greater) of the sets $f^{-1}(0)$ and $f^{-1}(1)$, and let the set system

$$\mathcal{F}_y = \big\{ \{x\} \mid x \in S,\ x \neq y \big\},$$

and let furthermore

$$\mathcal{F}'_y = \mathcal{F}_y \cup \mathcal{F}_y^- .$$

It is easy to see that $(S, \mathcal{F}'_y, f)$ is a recognition problem and

$$g(S, \mathcal{F}'_y, f) = \max\big(|f^{-1}(0)|, |f^{-1}(1)|\big) \geq \frac{|S|}{2}.$$

It is also easy to choose a non-constant function f, whose complexity is small over a given set system $\mathcal{R}$: set $f^{-1}(0)$ to be an arbitrary element of $\mathcal{R}$, then $g(S,\mathcal{R},f) = 1$. However, when we are trying to find a function, whose complexity is large over a given set system $\mathcal{R}$, some difficulties arise. Not only the creation of such a function is difficult, but also to determine its complexity. Although Theorem 3.7. and Corollaries 3.8. and 3.9. give lower bounds on the complexity of an arbitrary function, so the maximum of these lower bounds considering all possible functions is obviously a lower bound on $\max_f g(S,\mathcal{R},f)$, these are not at all easy to compute in general, since the structure of $\mathcal{R}^\cap$ can be quite complicated. The following theorem gives a different approach instead. First the *density* of a set system is defined.

Definition 3.10. The *density* of a set system $\mathcal{R}$ on the underlying set S is defined to be $\frac{\log|\mathcal{R}|}{|S|}$, and is denoted by $\varrho_{\mathcal{R}}$.

Remark. At first glance it would be more logical to define density as $\frac{|\mathcal{R}|}{2^{|S|}}$, but from a search theoretical point of view, a set system of (say) $2^{|S|-3}$ elements is rated to be fairly large, thus it would be unwise to define its density to be $\frac{1}{8}$; the number $\frac{|S|-3}{|S|}$ is much more informative.

Theorem 3.11.

$$\max_f g(S, \mathcal{R}, f) > \log \frac{1}{\varrho_{\mathcal{R}}}$$

Proof. We show that there exists a function f, for which $\log 1/\varrho_{\mathcal{R}}$ questions are not enough. To every (successful) adaptive algorithm of k steps a search tree of depth k can be associated in the following way. The inner nodes are labelled with the questions and the leaves are labelled with the outcomes: 0 or 1. The number of distinct adaptive algorithms of k steps is therefore at most the number of such labelled trees, which is

$$\left(\frac{|\mathcal{R}|}{2}\right)^{2^k-1} \cdot 2^{2^k},$$

since the number of inner nodes and leaves are $2^k - 1$ and 2^k, respectively, and the number of possible questions is $|\mathcal{R}|/2$ for every inner node (note that $\mathcal{R}$ is closed under complementation and a set and its complement gives the same question). Thus if the number of different functions $2^{|S|}$ is greater than $\left(\frac{|\mathcal{R}|}{2}\right)^{2^k-1} \cdot 2^{2^k}$, then there exists two distinct functions, for which all the questions and outcomes are exactly the same, that is impossible. This shows that if

$$\left(\frac{|\mathcal{R}|}{2}\right)^{2^k-1} \cdot 2^{2^k} < 2^{|S|},$$

then k questions are not enough. Since this inequality holds for $k = \log 1/\varrho_{\mathcal{R}}$, the theorem is proved. ■

4. The generalized communication model

In the previous section we have seen several bounds that are concerning all test families satisfying conditions (C) and (M).

If the underlying set S is the set of the positions of a given matrix M and the set system $\mathcal{R}$ of the question sets is the set of all subsets of S consisting of either some rows or some columns of M, then we obtain a model which is equivalent to two-party deterministic communication complexity [13]. This model can be generalized such that some communication complexity theorems remain valid [13]. In this section we summarize some theorems concerning the connection between recognition and communication complexity. For details see [13].

4.1. The communication model

First we prove that two-party deterministic communication complexity is indeed a special case of recognition complexity.

Consider a given communication complexity problem with input sets A and B for Alice and Bob, respectively, and a function $f : A \times B \longrightarrow \{0, 1\}$.

The communication problem can be reformulated as a recognition problem in the following way: let the underlying set be $S = A \times B$ and the set system of the question sets be

$$\mathcal{R} = \{U \times A \mid U \subseteq B\} \cup \{T \times B \mid T \subseteq A\}.$$

Since conditions (C) and (M) hold, $(S, \mathcal{R}, f)$ is indeed a recognition problem.

The sequence of bits communicated between Alice and Bob is the same as we get from the answers of the recognition problem supposing that we choose the question set $T \times B$ when Alice gives a bit information about set $T \subseteq A$, and the question set $U \times A$ when Bob gives a bit information about $U \subseteq B$. Furthermore, the communication and the recognition problems terminate at the same time: when the set of the still possible elements becomes monochromatic. Now it is easy to see that

$$g(S, \mathcal{R}, f) = D(f).$$

Remark. Various models of more than two party communication can also be treated as recognition problems and also can be generalized in this way.

Considering this relationship between recognition and communication problems, the idea of applying communication complexity methods to recognition problems arises quite naturally. We have seen some applications of this type, like Theorem 3.7. or Corollary 3.9. In order to be able to apply more methods of communication complexity, we have to give some restrictions on the set system of the question sets, since this system is quite special in the communication model. How should such a restriction look like? Some of the methods of communication complexity is based on the fact that $\mathcal{R}^{\cap} = \mathcal{R}^{2\cap}$ (since both set systems are the set of rectangles) in the communication model. Assuming that this equality holds for a recognition problem $(S, \mathcal{R}, f)$, some communication type bounds can be proved, indeed. Let us demonstrate this.

First we define partition and cover numbers. These numbers are generalizations of the partition and cover numbers of communication complexity (see [7]).

Definition 4.1. Let $(S, \mathcal{R}, f)$ be a recognition problem.

1. The *cover number* of f, denoted by $C(f)$ is the minimal number of f-monochromatic sets belonging to $\mathcal{R}^{\cap}$ that cover the underlying set S.
2. The *partition number* of f, denoted by $P(f)$ is the minimal number of f-monochromatic sets belonging to $\mathcal{R}^{\cap}$ that partition the underlying set S.
3. The *algorithmic partition number* of f, denoted by $A(f)$ is the minimal number of leaves of the decision trees corresponding to a successful recognition algorithm for the function f.
4. The minimal number of f-monochromatic sets of $\mathcal{R}^{\cap}$ that cover all elements x, for which $f(x) = z$ is denoted by $C^z(f)$ $\big(z \in \{0,1\}\big)$.

The following proposition summarizes the obvious properties of these numbers.

Proposition 4.2.

$$C^0(f) + C^1(f) = C(f) \leq P(f) \leq A(f) \leq 2^{g(f)}.$$

Proof. The equality and the first two inequalities are straightforward. Consider an algorithm of length $g(f)$ for the function f. The corresponding decision tree has at most $2^{g(f)}$ leaves, which implies the third inequality. ■

Remark. Theorem 3.7. stated $P(f) \leq 2^{g(f)}$.

Now we give a generalization of a well-known bound of communication complexity.

Proposition 4.3. *Let $(S, \mathcal{R}, f)$ be a recognition problem, for which $\mathcal{R}^{\cap} = \mathcal{R}^{2\cap}$. Then*

$$\log A(f) \leq g(f) \leq 3 \log A(f).$$

Proof. The first inequality follows from Proposition 4.2. In order to prove the second one, consider a node $\mathbf{N}$ of the search tree for which the number $D(\mathbf{N})$ of leaves that are descendants of $\mathbf{N}$ is at least half of the number $A(f)$ of all leaves, and is minimal amongst the nodes satisfying this property. Such a node always exists. Denote the sons of $\mathbf{N}$ by $\mathbf{K}$ and $\mathbf{M}$, the appropriate sets (containing all elements of S that still can be equal to the hidden x at the current node) by N, K, and M, respectively. The numbers $D(\mathbf{K})$ and $D(\mathbf{M})$ are smaller than $D(\mathbf{N})$, thus $D(\mathbf{K}) \leq \frac{A(f)}{2}$ and $D(\mathbf{M}) \leq \frac{A(f)}{2}$. Furthermore, the number of leaves that are not descendants of $\mathbf{N}$ is also at most $\frac{A(f)}{2}$. The set N at node **N** is of course a member of $\mathcal{R}^{\cap}$ so, since $\mathcal{R}^{\cap} = \mathcal{R}^{2\cap}$, there exist sets $X, Y \in \mathcal{R} : N = X \cap Y$. Now we change the search tree in the following way. Let the next two questions be "is $x \in X$?" and "is $x \in Y$?". If at least one of the sets X and Y does not contain x, then we delete that part of the tree which contains descendants of **N**, whilst if both X and Y contain x, then we ask one more question: the same as we would ask at node **N** in the recognition problem corresponding to the original search tree. Depending on the answer, we consider either the **K**- or the **M**-rooted subtree and delete everything else. In this way, using at most three questions, we obtain a tree, which has at most half as many leaves as the original one. Iterating the process we obtain an algorithm of at most $3 \log A(f)$ steps. ■

The following proposition is also a generalization of a communication complexity upper bound.

Proposition 4.4. *Let $(S, \mathcal{R}, f)$ be a recognition problem, for which $\mathcal{R}^{\cap} = \mathcal{R}^{2\cap}$. Then*

$$g(f) \le C^z(f) + 2, \quad \text{for} \quad z \in \{0, 1\}.$$

Proof. We give a nearly static algorithm of length $C^z(f) + 2$. Let

$$\left\{R_1, R_2, \ldots, R_{C^z(f)}\right\} \subseteq \mathcal{R}^{\cap}$$

be a z-monochromatic cover of the elements of $f^{-1}(z)$. Since $\mathcal{R}^{\cap} = \mathcal{R}^{2\cap}$, $R_i = A_i \cap B_i$, where $A_i, B_i \in \mathcal{R}$. The first $C^z(f)$ questions of the algorithm are "$x \in A_i$?" for $i = 1, 2, \ldots, C^z(f)$. Now let I be the set of those indices i, for which $x \in A_i$, and let

$$C = \bigcup_{i \in \mathrm{I}} B_i.$$

$C \in \mathcal{R}^{2\cup} = \mathcal{R}^{\cup}$ (this follows from $\mathcal{R}^{\cap} = \mathcal{R}^{2\cap}$ and condition (C)), thus $C = C_1 \cup C_2$, for some $C_1, C_2 \in \mathcal{R}$. We know the index set I, thus we can determine C, and then C_1 and C_2. Let the last two questions be "$x \in C_1$?" and "$x \in C_2$?". If the hidden element x belongs to at least one of C_1 and C_2, then $f(x) = z$, otherwise $f(x) = \overline{z}$. To prove this, consider $C = C_1 \cup C_2$. If $x \in C$, then there exists an index i such that both A_i and B_i contain x, thus $f(x)$ must be z, because $A_i \cap B_i$ is monochromatic. On the other hand, if $x \notin C$, then none of the sets R_i contain x, thus $f(x)$ cannot be z, since $R_1, R_2, \ldots, R_{C^z(f)}$ is a cover of all elements on which f gives the value z. ■

Finally we mention a (not too strong) property of coordinates if the condition $\mathcal{R}^{\cap} = \mathcal{R}^{2\cap}$ holds.

Proposition 4.5. *If an element x has at least two coordinates in a set system, for which $\mathcal{R}^{\cap} = \mathcal{R}^{2\cap}$, then it has two coordinates whose intersection is $\{x\}$.*

Proof. Obvious, by Proposition 3.6. ∎

We have demonstrated that the condition $\mathcal{R}^{\cap} = \mathcal{R}^{2\cap}$ can be used to generalize certain communication complexity bounds. However, this condition is not strong enough. For example, concerning Proposition 4.5., an upper bound on the number of the coordinates of an element would be better. Another problem is that this condition is not "algorithmic" in the sense that though there exist two sets, whose intersection is a given set $A = A_1 \cap A_2 \cap A_3 \cap \ldots \in \mathcal{R}^{\cap}$, these sets have nothing in common with the sets $A_1, A_2, A_3, \ldots$.

Remark. Assuming $\mathcal{R}^{\cap} = \mathcal{R}^{k\cap}$ for some $k \geq 3$ instead of $\mathcal{R}^{\cap} = \mathcal{R}^{2\cap}$, we obtain similar (of course somewhat weaker) results.

4.2. Quadrangular set systems

Now we give a condition that is stronger than the condition we have seen in the previous section.

Definition 4.6. Set systems $\mathcal{A}$ satisfying the following condition (F) are called *quadrangular set systems.*

$$\text{(F)} \qquad A, B, C \in \mathcal{A} \Rightarrow (A \cap B \in \mathcal{A}) \vee (A \cap C \in \mathcal{A}) \vee (B \cap C \in \mathcal{A}).$$

It can be readily seen that if condition (C) holds, then (F) is equivalent to the condition

$$\text{(F}'\text{)} \qquad A, B, C \in \mathcal{A} \Rightarrow (A \cup B \in \mathcal{A}) \vee (A \cup C \in \mathcal{A}) \vee (B \cup C \in \mathcal{A}).$$

Definition 4.7. If $(S, \mathcal{A}, f)$ is a recognition problem, where set the system $\mathcal{A}$ is quadrangular, then $(S, \mathcal{A}, f)$ itself is called quadrangular.

It is easy to see that the communication model is quadrangular, since among any three question sets there are at least two of the same type (either row or column sets), whose intersection and union are also row or column sets. The name "quadrangular" comes from this fact, since the intersection of row and column sets are called rectangles. We could also name these systems about their correspondence to alignments (set systems that are closed under intersection); they are "almost" alignments. Now we prove that condition (F) is stronger than the condition $\mathcal{A}^{\cap} = \mathcal{A}^{2\cap}$ indeed.

Proposition 4.8. *Condition* (F) *implies* $\mathcal{A}^{\cap} = \mathcal{A}^{2\cap}$.

Proof. $\mathcal{A}^{2\cap} \subseteq \mathcal{A}^{\cap}$ is trivial. Let $C = A_1 \cap A_2 \cap \ldots \cap A_r \in \mathcal{A}^{\cap}$. For $r \leq 2$, $C \in \mathcal{A}^{2\cap}$. For $r \geq 3$, at least one of the sets $A_1 \cap A_2$, $A_1 \cap A_3$, and $A_2 \cap A_3$ belongs to $\mathcal{A}$, by condition (F), thus $\mathcal{A}^{\cap} \subseteq \mathcal{A}^{2\cap}$ follows by induction. ■

Besides the communication model, examples of quadrangular set systems can be found in [13]. Some of these examples will be examined in details in the next section.

Further on we assume that the set system $\mathcal{A}$ satisfies conditions (C), (M), and (F).

Once we know that the hidden element x lies in a set $X \subseteq S$, we can try to simplify the search problem by taking X as an underlying set instead of S. It is possible if and only if the set system

$$\mathcal{A}|_X = \{A \cap X \mid A \in \mathcal{A}\}$$

satisfies conditions (C) and (F) (condition (M) is trivially satisfied). Fortunately it is always true:

Lemma 4.9. *Let $X \subseteq S$ be an arbitrary set. Then the set system $\mathcal{A}|_X$ satisfies conditions (C) and (F).*

Proof. Let $C \cap X$, $D \cap X \in \mathcal{A}|_X$. Then

$$X \setminus (C \cap X) = \overline{C} \cap X \in \mathcal{A}|_X \Rightarrow (C),$$

$$(C \cap X) \cap (D \cap X) = (C \cap D) \cap X \in \mathcal{A}|_X \Rightarrow (F). \qquad \blacksquare$$

Now we prove a property that is "natural" in two-party communication complexity.

Lemma 4.10. *Every element $x \in S$ has at most two coordinates.*

Proof. Assume to the contrary that there exists an element x with three distinct coordinates A, B, and C. Then by condition (F), (at least) one of the sets $A \cap B$, $A \cap C$, and $B \cap C$, each containing x, belongs to $\mathcal{A}$, which contradicts the minimality of A, B, and C. ∎

Definition 4.11. The number of distinct coordinates in $\mathcal{A}$ is denoted by $C(\mathcal{A})$.

The following theorem shows that the communication model is extremal amongst quadrangular problems in a certain sense, namely the number of distinct coordinates is minimum for the communication model, especially when the communication matrix is squared: $C(\mathcal{A}) = 2\sqrt{|S|}$ then.

Theorem 4.12. *Let $\mathcal{A} \subseteq 2^S$ be a quadrangular system. Then*

$$C(\mathcal{A}) \geq 2\sqrt{|S|}.$$

The proof of this theorem can be found in [13], just like the proof of the following lower bound, which can be obtained from a completely different approach.

Proposition 4.13. $C(\mathcal{A}) \geq \log |\mathcal{A}|$.

5. Recognition problems in partially ordered sets

In this section we analyze some special quadrangular problems concerning partially ordered sets.

Let S be an arbitrary finite underlying set. Let the set system $\mathcal{T} \subseteq 2^S$ be closed under union and intersection. Now let the set system $\mathcal{A}$ consist of sets of $\mathcal{T}$ and their complements:

$$\mathcal{A} = \mathcal{T} \cup \mathcal{T}^{-}.$$

Condition (C) is obviously satisfied, and the same holds for condition (F): among any three sets at least two belong to either $\mathcal{T}$ or $\mathcal{T}^{-}$. Since not only $\mathcal{T}$, but also $\mathcal{T}^{-}$ is closed under union and intersection, by the DeMorgan's laws, the intersection of these two sets also belongs to $\mathcal{A}$. Condition (M) is not necessarily satisfied, but if we contract the minimal elements of $\mathcal{A}^{\cap}$ to one element we obtain a set system satisfying all three conditions.

An example for a set system of this type is easy to give: let $P = (S, \preceq)$ be an arbitrary poset and let $\mathcal{T}$ be the set system of all downsets in the partial order $\preceq$. (A set $T \subseteq S$ is a *downset* (*upset*), if $x \in T$ implies $y \in T$ for every $y \preceq x$ ($x \preceq y$).) It can be readily seen that $\mathcal{T}$ is closed under union and intersection indeed. In fact, every set system of this type can be obtained in this way:

Proposition 5.1. *Let $\mathcal{T} \subseteq 2^S$ be a set system which covers S and is closed under union and intersection and suppose furthermore that condition (M) holds for $\mathcal{T}$. Then there exists a partial order $\preceq$ on S, such that the set system containing all downsets in $(S, \preceq)$ is $\mathcal{T}$.*

Proof. First we give a relation $\preceq$ on $S \times S$, then we show that $\preceq$ is a partial order, and finally prove that the system of all downsets in $(S, \preceq)$ is precisely the set system $\mathcal{T}$. Let

$$x \preceq y \quad \text{if} \quad x \in \bigcap_{\substack{y \in Y \\ Y \in \mathcal{T}}} Y.$$

It is obvious that $\preceq$ is reflexive. Transitivity is also easy to see: if $a \preceq b \preceq c$, then a is contained in every set of $\mathcal{T}$ that contains b, and b is contained in every set of $\mathcal{T}$ that contains c, thus a is contained in every set of $\mathcal{T}$ that contains c, so $a \preceq c$. Antisymmetry follows from condition (M): if $a \preceq b$ and $b \preceq a$, then a and b belong to the same sets of $\mathcal{T}$, thus $a = b$, by (M). Thus we have shown that $\preceq$ is a partial order.

Now let $T \in \mathcal{T}$. We show that T is a downset, i.e. if $t \in T$ and $s \preceq t$, then $s \in T$. The relation $s \preceq t$ implies that every set of $\mathcal{T}$ containing t also contains s. Thus T must contain s. On the other hand, we have to see that if a set $T \in 2^S$ is a downset, then $T \in \mathcal{T}$. To see this, let T' be the set of the maximal elements of T in the partial order $\preceq$. Since T is a downset,

$$T = \{x \mid \exists y \in T' : x \preceq y\} = \bigcup_{y \in T'} \{x \mid x \preceq y\}.$$

Since $\mathcal{T}$ is closed under union, it is enough to show that the sets $\{x \mid x \preceq y\}$ belong to $\mathcal{T}$. By the definition of the partial order $\preceq$,

$$\{x \mid x \preceq y\} = \bigcap_{\substack{y \in Y \\ Y \in \mathcal{T}}} Y.$$

Since $\mathcal{T}$ covers S, this intersection is not empty, and since $\mathcal{T}$ is closed under intersection, $\{x \mid x \preceq y\}$ belongs to $\mathcal{T}$ for every y, which completes the proof of the proposition. ∎

Now we examine recognition problems using question sets of this type. Similar studies can be found in [13], however, the results presented here are slightly stronger.

Let $(S, \preceq)$ be an arbitrary partially ordered set and $f : S \to \{0, 1\}$ be an arbitrary Boolean function. Denote the system of all up- and downsets in $(S, \preceq)$ by $\mathcal{S}_{\preceq}$. Then we would like to determine $g(S, \mathcal{S}_{\preceq}, f)$. The question "does x belong to $A \in \mathcal{S}_{\preceq}$?" is clearly equivalent to the question "what is $a(x)$?", where function $a : S \to \{0, 1\}$ is given by

$$a(y) = \begin{cases} 1, & \text{if } y \in A \\ 0, & \text{if } y \notin A. \end{cases}$$

Observe that the Boolean function a is monotonically decreasing (i.e. $x \preceq y$ implies $a(y) \le a(x)$) if and only if the set A is a downset and monotonically increasing (i.e. $x \preceq y$ implies $a(x) \le a(y)$) if and only if A is an upset. Thus the information we obtain from a question is precisely the value of an arbitrary monotone function on the hidden element x. From now on we shall also use monotone Boolean functions as questions.

As a warm up, consider the set of 0–1 vectors of length n and their standard partial order ($x \preceq y$ if for every index i and for the coordinates x_i of x and y_i of y we have $x_i \le y_i$). The number of 1 coordinates of a vector x will be called the length of x. It is easy to see that $\lceil \log(n+1) \rceil + 1$ questions are always enough[1]. First we find out the number of ones in the hidden vector with binary search. The set of vectors having length at most t is a downset for every t, so let the first question set be $\{x \mid x$ has length at most $\lfloor \frac{n-1}{2} \rfloor\}$. Obtaining the answer we ask either the set $\{x \mid x$ has length at most $\lfloor \frac{n-1}{4} \rfloor\}$ (if the first answer was "yes") or the set $\{x \mid x$ has length at most $\lfloor \frac{3n-1}{4} \rfloor\}$ (if the first answer was "no"), and so on. In this way using $\lceil \log(n+1) \rceil$ questions (the length may be any integer between 0 and n) we learn the length of the hidden vector x.

After the length of x proved to be some number k, we only have to ask one further question: function $a : S \to \{0, 1\}$, defined in the following way:

$$a(y) = \begin{cases} 1, & \text{if } y \text{ contains more than } k \text{ ones} \\ f(y), & \text{if } y \text{ contains exactly } k \text{ ones} \\ 0, & \text{if } y \text{ contains less than } k \text{ ones.} \end{cases}$$

[1]Later on we shall see an algorithm that uses at most $\lceil \log(n+1) \rceil$ questions.

It is obvious that a is monotone and since we know that vector x contains exactly k ones, $f(x) = a(x)$.

Of course it is possible that we do not need so many questions. For example, if f itself is monotone, we may simply ask $f(x)$. Now we show that there exists a function f, for which the above algorithm is nearly optimal. Observe that the elements of $\mathcal{S}_{\preceq}^{\cap}$ are precisely the convex sets of the poset $(S, \preceq)$, i.e. sets C with the following property:

$$(w, y \in C,\ w \preceq y,\ w \preceq z \preceq y) \Rightarrow z \in C.$$

We define the function f on the vector x to be the parity of the length of x. Now we apply Theorem 3.7. The set of vectors clearly cannot be partitioned into less than $(n+1)$ f-monochromatic convex sets, hence $g(f) \geq \lceil \log(n+1) \rceil$.

Now we turn our attention to the general case: what is the exact number of monotone functions we have to ask for a given function f? In order to answer this, we shall generalize the above ideas. First we define the *alternation number* of a chain and a function.

Definition 5.2. Let $(S, \preceq)$ be an arbitrary poset, $f : S \to \{0,1\}$ be an arbitrary Boolean function.

A chain D is said to be *f-alternating,* if $x, y \in D$, $x \preceq y$, $\nexists z \in D$: $x \preceq z \preceq y$ implies $f(x) \neq f(y)$.

The *alternation number* of an arbitrary chain C in $(S, \preceq)$ is the size of the (inclusionwise) maximal f-alternating subchain of C. The alternation number of C is denoted by $\mathrm{Alt}_P(C)$ for a given $P = (S, \preceq)$ and f.

The *alternation number of a function f* is

$$\mathrm{Alt}_P(f) = \max_{\substack{C \subseteq S \\ \text{chain}}} \mathrm{Alt}_P(C).$$

The following theorem essentially determines the complexity of the recognition problem $(S, \mathcal{S}_{\preceq}, f)$.

Theorem 5.3.

$$\lceil \log \mathrm{Alt}_P(f) \rceil \leq g(f) \leq \lceil \log \mathrm{Alt}_P(f) \rceil + 1.$$

Proof. After a question is answered (say $a(x) = 1$) we can work with the poset $(S \cap a^{-1}(1), \preceq)$, instead of the poset $(S, \preceq)$, by Lemma 4.9. and the observation preceding it.

First we prove $g(f) \geq \lceil \log \mathrm{Alt}_P(f) \rceil$. Let C be a chain with the property $\mathrm{Alt}_P(C) = \mathrm{Alt}_P(f)$. The elements of the (inclusionwise) maximal f-alternating subchains form a fooling set of cardinality $\mathrm{Alt}_P(f)$, thus by Corollary 3.9. $g(f) \geq \lceil \log \mathrm{Alt}_P(f) \rceil$.

Now we prove the second part of the inequality: $g(f) \leq \lceil \log \mathrm{Alt}_P(f) \rceil + 1$.

If $\mathrm{Alt}_P(f) = 1$, then one question is enough, because f must be monotone on the set of the elements that are still candidates (it is not sure however that f is constant, because the poset need not to be connected). Thus it is enough to prove that it is possible to decrease the value $\lceil \log \mathrm{Alt}(f) \rceil$ by one with one question: from this the inequality follows by induction.

Let us denote the set $\{y \in S \mid y \preceq z\}$ by S_z, and the poset $(S_z, \preceq)$ by P_z. Now let

$$A = \left\{ z \mid \mathrm{Alt}_{P_z}(f) \leq \frac{\mathrm{Alt}_P(f)}{2} \right\}.$$

A is a downset, otherwise we could choose $z \preceq v$, so that

$$z \notin A, \quad v \in A,$$

thus

$$\mathrm{Alt}_{P_z}(f) > \frac{\mathrm{Alt}_P(f)}{2}, \quad \mathrm{Alt}_{P_v}(f) \leq \frac{\mathrm{Alt}_P(f)}{2},$$

from which

$$\mathrm{Alt}_{P_z}(f) > \mathrm{Alt}_{P_v}(f),$$

a contradiction, since clearly $S_z \subseteq S_v$, thus $\mathrm{Alt}_{P_z}(f) \leq \mathrm{Alt}_{P_v}(f)$.

We have seen that the set A is a downset, therefore if

$$\text{(1)} \qquad \mathrm{Alt}_{(A,\preceq)}(f) \leq \left\lceil \frac{\mathrm{Alt}_P(f)}{2} \right\rceil \quad \text{and} \quad \mathrm{Alt}_{(S \setminus A,\preceq)}(f) \leq \left\lceil \frac{\mathrm{Alt}_P(f)}{2} \right\rceil,$$

then it is possible to decrease the value $\lceil \log \mathrm{Alt}(f) \rceil$ by one with one question, namely by asking the set A.

Now we show that (1) holds. Let us assume to the contrary that there exists a chain C, such that

$$\text{(2)} \qquad C \subseteq A \quad \text{and} \quad \mathrm{Alt}_{(A,\preceq)}(C) > \left\lceil \frac{\mathrm{Alt}_P(f)}{2} \right\rceil$$

or

$$(3) \qquad C \subseteq S \setminus A \quad \text{and} \quad \mathrm{Alt}_{(S\setminus A,\preceq)}(C) > \left\lceil \frac{\mathrm{Alt}_P(f)}{2} \right\rceil .$$

Let furthermore C be maximal amongst those chains that satisfy (2) or (3). (2) clearly contradicts the definition of A, while if (3) holds, then we show that the minimal element m of the chain C cannot be contained in $S \setminus A$, again a contradiction. Namely, if m were an element of $S \setminus A$, then

$$\mathrm{Alt}_{P_m}(f) > \frac{\mathrm{Alt}_P(f)}{2} \Rightarrow \exists \text{ chain } C' \subseteq S_m : \mathrm{Alt}_{P_m}(C') > \frac{\mathrm{Alt}_P(f)}{2}.$$

Consider now the chain $C'' = C' \setminus \{m\}$. $C'' \subseteq A$, since m is the minimal element of the maximal chain C and clearly

$$(4) \qquad \mathrm{Alt}_{(A,\preceq)}(C'') > \frac{\mathrm{Alt}_P(f)}{2} - 1.$$

Now by (3)

$$\mathrm{Alt}_P(C \cup C'') \geq \mathrm{Alt}_{(S\setminus A,\preceq)}(C) + \mathrm{Alt}_{(A,\preceq)}(C'') > \left\lceil \frac{\mathrm{Alt}_P(f)}{2} \right\rceil + \mathrm{Alt}_{(A,\preceq)}(C''),$$

thus by (4)

$$\mathrm{Alt}_P(C \cup C'') \geq \left\lceil \frac{\mathrm{Alt}_P(f)}{2} \right\rceil + 1 + \mathrm{Alt}_{(A,\preceq)}(C'')$$

$$> \left\lceil \frac{\mathrm{Alt}_P(f)}{2} \right\rceil + 1 + \frac{\mathrm{Alt}_P(f)}{2} - 1 = \left\lceil \frac{\mathrm{Alt}_P(f)}{2} \right\rceil + \frac{\mathrm{Alt}_P(f)}{2} \geq \mathrm{Alt}_P(f),$$

a contradiction, since $C \cup C''$ is a chain in P. This completes the proof of the second inequality and thus the whole theorem. ■

The number $\mathrm{Alt}_{P_z}(f)$ may be called the *level* of z (for function f). Observe that if either $\mathrm{Alt}_{(A\cap S_z,\preceq)}(f) = \mathrm{Alt}_{(A\cap S_y,\preceq)}(f)$ or $\mathrm{Alt}_{(\overline{A}\cap S_z,\preceq)}(f) = \mathrm{Alt}_{(\overline{A}\cap S_y,\preceq)}(f)$, then $\mathrm{Alt}_{(S_z,\preceq)}(f) = \mathrm{Alt}_{(S_y,\preceq)}(f)$, that is, y and z are in the same level. Therefore, by induction we obtain that the elements of the final poset Z (for which we have $\mathrm{Alt}_Z(f) = 1$) are in the same level. This may be useful, because if the set of the minimal elements of $P = (S, \preceq)$ is monochromatic, then the sets of elements of the same level are also monochromatic, thus the final poset is monochromatic, too. This means that the last question is unnecessary, and therefore in these cases the recognition complexity is precisely $\lceil \log \mathrm{Alt}_P(f) \rceil$. This holds automatically when a unique minimal element exists (for example, in the case of 0–1 sequences, proving that the complexity is $\lceil \log(n+1) \rceil$ in the warm up problem). Obviously, if a unique maximal element exists, it is also enough, by considering the dual of the poset. Naturally arises the idea that if there is no unique minimal (nor unique maximal) element exists, then we may create one. Let $P = (S, \preceq)$ be an arbitrary poset. Consider now poset $P^0 = \big(S \cup \{\mathbf{0}\}, \preceq^0\big)$, where $\mathbf{0} \notin S$ and $\preceq^0 = \preceq \cup \bigcup_{s \in S}(\mathbf{0}, s)$. We have seen that

$$g(S \cup \mathbf{0}, P^0, f) = \lceil \log \mathrm{Alt}_{P^0}(f) \rceil.$$

If we can determine $f(x)$ for the new poset, then we can determine it to the original one obviously. $\mathrm{Alt}_{P^0}(f) \leq \mathrm{Alt}_P(f) + 1$, so we have proved a stronger version of Theorem 5.3.:

Theorem 5.4.

$$\lceil \log \mathrm{Alt}_P(f) \rceil \leq g(f) \leq \lceil \log\big(\mathrm{Alt}_P(f) + 1\big) \rceil. \qquad \blacksquare$$

Finally in this section, we are dealing with a special poset P: graphs with their standard partial order ($G_1 \preceq G_2$ if the two graphs have the same vertex-set and every edge of G_1 is an edge of G_2, that is, if G_1 is a spanning subgraph of G_2). A graph property T is given on graphs having n vertices (i.e. $T : \mathcal{G}_n \to \{0,1\}$ is a function with the additional condition that $T(G_1) = T(G_2)$, if graphs G_1 and G_2 are isomorphic, where $\mathcal{G}_n$ is the set of all graphs having n vertices); our aim is to determine $T(G)$ for an unknown graph G. If we are allowed to ask any monotone functions, then we can simply apply the theorem we have just proved. However, it seems to be more logical if we are allowed to ask monotone graph properties only. The lower bound obviously remains true, the question is what happens to the upper bound.

Let $\mathcal{F}_n$ be the set of all monotone graph-properties on $\mathcal{G}_n$.

Remark. The name "monotone graph property" usually refers to monotonically increasing graph-properties, and monotonically decreasing properties are called "co-monotone properties". For the sake of consistency, we use the same names as for functions, since properties are functions by our definition. Thus monotone means monotonically increasing or monotonically decreasing.

Let furthermore

$$\mathcal{R}_n = \left\{T^{-1}(0) \mid T \in \mathcal{F}_n\right\} \cup \left\{T^{-1}(1) \mid T \in \mathcal{F}_n\right\}.$$

Theorem 5.5.

$$g(\mathcal{G}_n, \mathcal{R}_n, T) = \left\lceil \log \mathrm{Alt}_{(\mathcal{G}_n, \preceq)}(T) \right\rceil.$$

Proof. Consider the poset $P^* = (\mathcal{G}_n^*, \preceq^*)$ of isomorphism classes of graphs on n vertices, where $G_1^* \preceq^* G_2^*$ holds if there exist graphs $G_1 \in G_1^*$ and $G_2 \in G_2^*$, such that G_1 is a subgraph of G_2. It is obvious that this relation is a partial order indeed and that

$$g(\mathcal{G}_n, \mathcal{R}_n, T) = g(\mathcal{G}_n^*, \mathcal{G}^*_{n\preceq^*}, T^*),$$

where T^* is a Boolean function on the isomorphism classes of graphs on n vertices given by $T^*(G^*) = T(G)$ for any $G \in G^*$.

Since the poset P^* has a unique minimal element,

$$g(\mathcal{G}_n^*, \mathcal{G}^*_{n\preceq^*}, T^*) = \left\lceil \log \mathrm{Alt}_{P^*}(T^*) \right\rceil.$$

Now it only remains to show that

$$\left\lceil \log \mathrm{Alt}_{P^*}(T^*) \right\rceil = \left\lceil \log \mathrm{Alt}_{(\mathcal{G}_n, \preceq)}(T) \right\rceil,$$

which is straightforward. ■

6. Predetermined complexity

Sometimes we would like to determine the value $f(x)$ for the hidden element x such that all questions are asked beforehand. In this case we speak about the *predetermined recognition complexity* of the function f. A special case we have seen is determining the simultaneous communication complexity of a function (see Section 2.2).

First we give the precise definition, then examine the problem for arbitrary set systems of question sets, and finally, as an application, compute the predetermined complexity of an arbitrary function for a set system we are quite familiar with: the up- and downsets of an arbitrary poset.

Definition 6.1. The *non-adaptive* or *predetermined recognition complexity* of a function f over the set system $\mathcal{R}$ on S is the smallest number of questions of type "is $x \in A$?" ($A \in \mathcal{R}$) needed to compute $f(x)$ in the worst case, where all questions are fixed beforehand. This number is denoted by $g_{\text{pre}}(S, \mathcal{R}, f)$, or simply by $g_{\text{pre}}(f)$, when S and $\mathcal{R}$ are fixed.

If the answers to the questions "is $x \in A_i$?" for $i = 1, 2, \ldots, r$ determine $f(x)$, then the system $\{A_1, A_2, \ldots, A_r\}$ is called a separating system with respect to f. So the predetermined complexity of f is the size of the smallest separating system with respect to f.

Proving that computing the simultaneous communication complexity of a function is a special case is similar to the proof of the fact that two-party deterministic communication complexity is indeed a special case of recognition complexity (see Section 4.1) and is omitted here. Though determining the simultaneous communication complexity is quite easy, the same is not true for the predetermined recognition complexity in general.

First we give an upper bound that is the generalization of an obvious upper bound for exact problems. However, we have to be careful with the proof, which is not so easy as it seems to be at the first sight.

Proposition 6.2.

$$g_{\text{pre}}(f) < A(f) \leq 2^{g(f)}.$$

Proof. The second inequality is a part of Proposition 4.2. To prove the first one, consider a decision tree D for function f having $A(f)$ leaves. To every inner node of the decision tree corresponds a question set. The system of all these question sets $\mathcal{A}$ form a separating system with respect to f, that is, the elements of

$$\min\left(\mathcal{A} \cup \mathcal{A}^-\right)^\cap$$

are monochromatic, otherwise there would exist a leaf v and $a, b \in S$, such that $f(a) = 1$, $f(b) = 0$, and both a and b reaches v, which is impossible. To see that such a leaf would exist indeed, let $T \in \min\left(\mathcal{A} \cup \mathcal{A}^-\right)^\cap$ be a non-monochromatic set. Since T is minimal in $\left(\mathcal{A} \cup \mathcal{A}^-\right)^\cap$, for any set $A \in \mathcal{A}$ either $T \subseteq A$ or $T \subseteq \overline{A}$.

Examine now the question set $A_1 \in \mathcal{A}$ corresponding to the root of the decision tree. We choose that son of the root that contains T. Since the sons of the root are A_1 and $\overline{A_1}$, and either $T \subseteq A_1$ or $T \subseteq \overline{A_1}$, we can do this. Now we choose the question set corresponding to the current node to be A_2, and choose that son of this node that contains T, and so on. Finally we obtain the leaf reached by all elements of T. Since T is not monochromatic, we have found a proper leaf.

Thus $\mathcal{A}$ is a separating system with respect to f indeed. Since $|\mathcal{A}|$ is at most the number of inner nodes of D, which is always smaller than the number of leaves of D, the proof is finished. ■

Next a lower bound using predetermined complexity of exact problems is given.

Proposition 6.3. *Let F be an arbitrary fooling set for function $f : S \to \{0,1\}$ on set system $\mathcal{R} \subseteq 2^S$. Then*

$$g_{\text{pre}}(S, \mathcal{R}, f) \geq L_{\text{pre}}\big(F, \mathcal{R}|_F\big).$$

Proof. Let $\mathcal{A} \subseteq \mathcal{R}$ be a minimal separating system with respect to f. Then the elements of $\min\left(\mathcal{A} \cup \mathcal{A}^-\right)^{\cap}$ are monochromatic (otherwise there would exist two elements a and b, such that $f(a) = 1,\ \ f(b) = 0$ and they belong to exactly the same sets of $\mathcal{A}$, clearly a contradiction).

We show that $\mathcal{A}|_F$ is a separating system on F, that is, if $a \neq b$ are arbitrary elements of F, then there exists a set $D \in \mathcal{A}|_F$ that separates them. Assume to the contrary that every set of $\mathcal{A}|_F$ contains either both of them or none of them. Then the same is true for every set of $\mathcal{A}$, of course, thus a and b belong to the same set of set system $\min\left(\mathcal{A} \cup \mathcal{A}^-\right)^{\cap}$. But this is impossible, since every set of $\min\left(\mathcal{A} \cup \mathcal{A}^-\right)^{\cap}$ is monochromatic and the elements a and b cannot belong to the same monochromatic set of $\mathcal{R}^{\cap}$, because they are (distinct) elements of fooling set F. (Obviously $\min\left(\mathcal{A} \cup \mathcal{A}^-\right)^{\cap} \subseteq \mathcal{R}^{\cap}$, since $\mathcal{A} \subseteq \mathcal{R}$ and $\mathcal{R}$ is closed under complementation.)

Thus

$$g_{\text{pre}}(S, \mathcal{R}, f) = |\mathcal{A}| \geq \big|\mathcal{A}|_F\big| \geq L_{\text{pre}}\big(F, \mathcal{R}|_F\big). \qquad ■$$

The above upper and lower bounds are generally not too close to each other. However, sometimes they are, and in these cases they serve very well to compute the predetermined complexity.

Proposition 6.4. *Let $\preceq$ be a partial order on the underlying set S, and let $\mathcal{S}_{\preceq}$ be the set system consisting of the up- and downsets of the partially ordered set $P = (S, \preceq)$. Let furthermore $f : S \to \{0,1\}$ be an arbitrary function. Then*

$$\text{Alt}_P(f) - 1 \leq g_{\text{pre}}(S, \mathcal{S}_{\preceq}, f) \leq \text{Alt}_P(f).$$

Proof. To obtain the lower bound we use Proposition 6.3. Let C be a chain in P, having alternation number $\mathrm{Alt}_P(f)$. Now let C' be a maximal f-alternating subchain of C. It is obvious that C' is a fooling set for f, thus

$$g_{\mathrm{pre}}(S, \mathcal{S}_{\preceq}, f) \geq L_{\mathrm{pre}}\big(C', \mathcal{S}_{\preceq}|_{C'}\big),$$

by Proposition 6.3.

Since C' is a chain,

$$\mathcal{S}_{\preceq}|_{C'} = \bigcup_{c \in C'} \big(\{f \in F \mid f \preceq c\} \cup \{f \in F \mid c \preceq f\}\big),$$

from which

$$L_{\mathrm{pre}}\big(C', \mathcal{S}_{\preceq}|_{C'}\big) = |C'| - 1 = \mathrm{Alt}_P(f) - 1$$

follows readily, and the lower bound is proved.

To obtain the upper bound, Proposition 6.2 is used. Consider an optimal adaptive algorithm for function f. The corresponding decision tree has at most $\mathrm{Alt}_P(f) + 1$ leaves, by the proof of Theorem 5.4. (from the proposition itself it only follows that the number of the leaves is at most $2^{\lceil \log(\mathrm{Alt}_P(f)+1) \rceil}$). Since

$$g_{\mathrm{pre}}(S, \mathcal{S}_{\preceq}, f) < A(f) \leq \mathrm{Alt}_P(f) + 1,$$

by Proposition 6.2., the upper bound is also proved. ■

7. Open problems

About the structure of quadrangular systems not much is known. However, we know a lot about alignments. Alignments are set systems closed under intersection, i.e. a set system $\mathcal{H} \subseteq 2^S$ is an *alignment* if $\mathcal{H} = \mathcal{H}^{\cap}$ (it is also required that $\emptyset, S \in \mathcal{H}$).

Quadrangular systems are "almost" alignments; can similar theorems be proved for them?

We may also be interested in further examples for quadrangular systems. Examples are not too difficult to give, but sequences of quadrangular systems different from the examples in [13] are not known.

The question whether there is a correspondence between $C(\mathcal{R})$ and $g(H, \mathcal{R}, f)$ is also open.

We have seen that for the poset problems, where $C(\mathcal{R})$ is the maximum possible, the recognition complexity is almost the same as the lower bound of Theorem 3.7., while for the communication model, where $C(\mathcal{R})$ is the minimum possible (Theorem 4.12.), this bound is not at all tight (see [7]).

Can we give better bounds if the alignment $\mathcal{R}^{\cap}$ is a special one, say a convex geometry? (An alignment $\mathcal{C}$ is said to be a *convex geometry* or an *anti-matroid* if it satisfies the so-called anti-exchange axiom: if $Y \in M$ and $x, y \in \overline{Y}$, then

$$y \in Y \vee x \quad \Rightarrow \quad x \notin Y \vee y,$$

where $A \vee b$ denotes the minimal set of $\mathcal{C}$ containing all elements of $A \cup \{b\}$ (this set exists uniquely, because $\mathcal{C}$ is an alignment). For details about alignments and convex geometries see for example [4] and [10].)

For example the set system $\mathcal{S}_{\preceq}^{\cap}$ is a convex geometry and the bounds given for $\mathcal{S}_{\preceq}^{\cap}$ are almost tight.

Acknowledgements. The author would like to thank Gyula O.H. Katona and Gábor Tardos for their suggestions.

REFERENCES

[1] M. Aigner, *Combinatorial Search,* John Wiley, Chichester and Teubner, Stuttgart (1988).

[2] B. Bollobás, *Extremal Graph Theory,* Academic Press (London – New York – San Francisco, 1978).

[3] D. Du and F. Hwang, *Combinatorial Group Testing,* World Scientific (1993).

[4] P. H. Edelman and R. E. Jamison, The theory of convex geometries, *Geometriae Dedicata,* **19** (1985), 247–270.

[5] G. Katona, Combinatorial search problems, in: *Survey of Combinatorial Theory* (ed. by J. Srivastava et al.), North-Holland (Amsterdam, 1973), pp. 285–308.

[6] G. Katona, Rényi and the combinatorial search problems, *Stud. Sci. Math. Hung.,* **26** (1991), 363–378.

[7] E. Kushilevitz and N. Nisan, *Communication Complexity,* Cambridge University Press (1997).

[8] T. Lengauer, VLSI theory, in: *Handbook of Theoretical Computer Science,* Vol. A, Elsevier (Amsterdam, 1990), pp. 835–868.

[9] L. Lovász, Communication complexity: a survey, in: *Paths, Flows and VLSI Layout* (B. H. Korte ed.), Springer Verlag (Berlin, 1990) (Early version in Tech. Rep. CS-TR-204-89, Princeton University, 1989).

[10] L. Lovász and M. Saks, Lattices, Möbius functions and communication complexity, *J. of Computer and System Sciences,* **47** (1993), 322–349. (Early version in *Proc. of 29th Annual Symposium on Foundations of Computer Science,* 1988, 81–90.)

[11] K. Mehlhorn and E. Schmidt, *Las Vegas is better then determinism in VLSI and distributed computing,* Proc. of 14th ACM Symposium on Theory of Computing (1982), pp. 330–337.

[12] A. Rényi, *Lectures on the theory of search,* Univ. North Carolina, Mimeo Series (1969).

[13] G. Wiener, Recognition problems and communication complexity, *Discrete Applied Mathematics,* **137(1)** (2004), 109–123.

[14] A. Yao, *Some complexity questions related to distributed computing,* Proc. of 11th ACM Symposium on Theory of Computing (1979), 209–213.

[15] H. P. Yap, *Some Topics in Graph Theory,* Cambridge University Press (Cambridge, 1986).

Gábor Wiener

Department of Computer Science and Information Theory
Budapest University of Technology and Economics

`wiener@cs.bme.hu`

and

Alfréd Rényi Institute of Mathematics
Hungarian Academy of Sciences

`wiener@renyi.hu`

MIX
Papier aus verantwortungsvollen Quellen
Paper from responsible sources
FSC® C105338

If you have any concerns about our products,
you can contact us on
ProductSafety@springernature.com

In case Publisher is established outside the EU,
the EU authorized representative is:
Springer Nature Customer Service Center GmbH
Europaplatz 3, 69115 Heidelberg, Germany

Printed by Libri Plureos GmbH
in Hamburg, Germany